Mesoporous Materials - Properties and Applications

Edited by Manjunath Krishnappa

Published in London, United Kingdom

IntechOpen

Supporting open minds since 2005

Mesoporous Materials - Properties and Applications
http://dx.doi.org/10.5772/intechopen.79068
Edited by Manjunath Krishnappa

Contributors
Ratna Ediati, Pramita Elfianuari, Eko Santoso, Dety Oktavia Sulistiono, Muhammad Nadjib, Hendrik Oktendy Lintang, Leny Yuliati, Yanan Gao, Lifeng Deng, Junfeng Zhang, Ranganatha S, Mohamed Ali Ghanem, Abdullah M. Al-Mayouf, Mabrook S. Amer, James Mgaya, Egid Mubofu

First published in London, United Kingdom, 2019 by IntechOpen
IntechOpen is the global imprint of INTECHOPEN LIMITED, registered in England and Wales, registration number: 11086078, The Shard, 25th floor, 32 London Bridge Street
London, SE19SG – United Kingdom
Printed in Croatia

British Library Cataloguing-in-Publication Data
A catalogue record for this book is available from the British Library

Additional hard and PDF copies can be obtained from orders@intechopen.com

Mesoporous Materials - Properties and Applications
Edited by Manjunath Krishnappa
p. cm.
Print ISBN 978-1-83880-649-1
Online ISBN 978-1-83880-650-7
eBook (PDF) ISBN 978-1-83880-650-7

For EU product safety concerns:
IN TECH d.o.o., Prolaz Marije Krucifikse Kozulić 3, 51000 Rijeka, Croatia, info@intechopen.com or visit our website at intechopen.com.

Meet the editor

Krishnappa Manjunath received his master's degree from Sri Venkateshwara University, Tirupathi, India. He received his PhD in Chemistry from Jain University, India, under the guidance of Prof. T. Ramakrishnappa in 2016. During his PhD, he was awarded a prestigious "CNPq-TWAS" fellowship in 2014 for a one-year internship and worked with Prof. Jairton Dupont, FRS, at the Laboratory of Molecular Catalysis, UFRGS, Brazil. In 2016, he joined the International Centre for Materials Science, JNCASR, Bangalore, where he is currently working as a research associate with Prof. C. N. R. Rao, FRS. His research interests mainly focus on investigations of aliovalent anion substitution in inorganic materials and their applications for water splitting reactions to generate hydrogen. He has authored more than 30 internationally reputed articles, mainly concerning advanced functional materials and nanomaterials for photocatalysis.

Contents

Preface

The basic theme of this book is to understand the fundamentals of porous functional materials, their properties, and significant applications. Mesoporous materials play a remarkable role in various applications, with their properties such as surface area, porosity, etc. In several fields, such as solar cells, batteries, photovoltaics, and energy conversions, mesoporous materials are considered to be important materials. This book covers the fundamentals of mesoporous materials, and various methods of synthesis, properties, and applications in different sectors. I hope that all the chapters in this book present advances and progress in the field of mesoporous materials.

Manjunath Krishnappa
Jawaharlal Nehru Centre for Advanced Scientific Research,
India

Section 1

Synthesis and Properties of Mesoporous Materials

Chapter 1

Synthesis and Characterizations of Titanium Tungstosilicate and Tungstophosphate Mesoporous Materials

Mohamed A. Ghanem, Abdullah M. Al-Mayouf and Mabrook S. Amer

Abstract

The work reports a development approach for the synthesis of novel multi-components mesoporous materials of titanium tungstate (*meso*-TiW) titanium tungstosilicate (*meso*-TiWSi) and tungstophosphate (*meso*-TiWP) mixed oxides that have high surface area and ordered mesoporous structures at nanometer length scale. Using the solvent evaporation-induced self-assembly (EISA) new oxides of bi- and tri-component of *meso*-TiW, *meso*-TiWSi and *meso*-TiWP oxides with different compositions and porosity were achieved. The physicochemical properties of the mesoporous oxides were characterized by X-ray diffraction, BET surface area analyzer, scanning, and transmission electron microscopes. Subject to the oxide composition, the obtained *meso*-TiW, *meso*-TiWSi and *meso*-TiWP exhibits high surface area, ordered 2D hexagonal mesostructured with order channels extended over a large area. The produced *meso*-TiW, *meso*-TiWSi, and *meso*-TiWP adsorbents exhibit good adsorption efficiency for the removal of Pb(II), Cd(II) and Hg(II) ions from water solution due to the presence of high surface area and accessibility of surface active sites. The adsorption efficiency of these mesoporous oxide reaches up to 95% and is found to be dependent contact time and adsorbents dose. The synthesis strategy is particularly advantageous for the production of new complex (multi-component) inorganic mesoporous materials that might have an application in the field of environmental, catalysis or energy storage and production.

Keywords: mesoporous, titanium tungstosilicate, tungstophosphate, surfactant self-assembly

1. Introduction

The approaches used for the synthesis of nanostructured materials show a critical role in synthesis the required materials and determining its final properties. To achieve mesoporous inorganic oxide powders with high surface area, uniformity and even mesoporous structures, there are many protocols have been implemented [1, 2] and the most important ones are: mesoporous silica materials as a host [3, 4], sol-gel precipitation, solution co-precipitation, thermal solid state, and surfactant self-assembled templates [5–8]. Generally, the sol-gel precipitation, co-precipitation, and surfactant self-assembled templates were the most widely used methods for preparation of mesoporous ion exchange materials.

The synthesis of nonsiliceous nanostructured mesoporous materials with high surface area using surfactant self-assembled templates has been recognized as a method for controlling the phase structure, ordering and pore size of these functional materials [9–12]. A wide range of mesoporous oxides, metals and semiconductors materials with regular geometries have been successfully prepared by direct templating from liquid crystalline phases [13]. Nevertheless the synthesis of nonsiliceous mesostructured still lags behind the synthesis of mesoporous silica materials, particularly for complex and multi-components inorganic materials. The reasons for this come from the complexity, reactivity, chemical stability and interaction of the surfactant with the inorganic precursor. Establishment of synthetic routes for multi-component metal oxide mesostructured those are effective, applicable, reproducible and suitable for large-scale production still remains a challenge.

Inorganic ion exchange materials are widely used for water treatment, in the semiconductor industry, and in nuclear technology. Multi-component ion-exchange materials, such as titanium tungstosilicate (TiWSi) and tungstophosphate (TiWP) mixed oxides have better chemical and thermal stability and exchange capacity compared to Ti (IV) tungstate, Ti (IV) silicate and Ti (IV) phosphate. The crystalline phases of these materials have been utilized for separating heavy metals in aqueous media. The adsorption, separation and exchange properties of these systems could potentially be enhanced significantly if the materials were obtained with mesoporous structures.

In this work, the solvent evaporation-induced self-assembly (EISA) template method is adapted to produce a selection of mesoporous ion exchange materials such as titanium tungstate (*meso*-TiW) titanium tungstosilicate (*meso*-TiWSi) and tungstophosphate (*meso*-TiWP) mixed oxides with the high surface area and controlled porous structures that similar to mesoporous silica. Using this method mesoporous ionic exchange materials can be synthesized with homogeneous and uniform structures in the form of powders, amenable to mass production, or as membranes depending on the conditions employed during the synthesis. The characterization and adsorption capacity of the resulting nanostructured mesoporous materials is investigated for water demineralization and the removal of heavy metal ions from water solution.

Using the template strategy to synthesize nanostructured mesoporous materials with a different shape, size and morphology was firstly introduced by Mobil scientists [14] and Martin *et al.* [15, 16]. In this template method, the desired materials are chemically formed using self-assembled surfactant templates (sol-gel) or electrochemically deposited inside the pores of a nanoporous membrane that subsequently dissolves away leaving a replica of nanomaterials structures. The templates are divided into two different types: hard and soft templates. The commonly used hard templates include anodic aluminum oxide [16], silica [17], and polystyrene spheres [18] and mesoporous carbon [19]. The soft templates are formed from surfactant molecules that are self-assembled into organized mesophases (soft matter) that are intermediate in order between crystalline solids and isotropic liquids.

In recent years, the sol-gel process is widely utilized concurrency with organic-directed assembly to synthesize a variety of non-siliceous mesostructured materials such as transition metals, metal oxides, semiconductors, carbons, and polymers. Such mesostructured materials have attracted significant attention because of the application of these materials in nanoscale devices, sensors, microelectronic devices, catalysts, energy storage media, and adsorbent materials. Mesoporous silica-based materials, for example, have been widely attracted an attention for heavy metal ions adsorption from wastewater [20–23]. This because they exceptionally have the large specific surface area, regular pore structure, and tunable pore sizes. Consequently, the mesoporous materials are expected to exhibit better heavy metal ions adsorption due to improved accessibility to the adsorption sites and/or speeding up the diffusion process within the mesopores. Toward this end and very recently, mesoporous silica materials were prepared by microwave assistant heating using silica fume as silica source and evaluated as adsorbents for the removal of Cu^{2+}, Pb^{2+}, and Cd^{2+} from aqueous solutions [24]. The results showed that the produced mesoporous silica could efficiently confiscate the heavy metal ions from aqueous solution within pH range of 5–7 and the maximum removal capacities of the mesoporous silica for Cu^{2+}, Pb^{2+}, and Cd^{2+} were 36.3, 58.5, and 32.3 mg/g, respectively.

In addition, mesoporous silica modified with suitable organic functional groups have shown enhanced heavy metals adsorbents capacity due to combining the open porous structure of mesoporous silica with suitable organic compounds that exhibit high binding affinity toward the target metal ions. As an example, Lee *et al.* [22] reported the adsorption of Pb(II) and Cu(II) metal ions on mesoporous silica modified with organic amine and thiol functional groups and results reveal that the thiol-functionalized mesoporous silica adsorbents exhibit enhanced adsorption affinity for Pb^{2+} than the amino-mesoporous silica. Moreover, other studies reported an excellent Hg removal efficiency on thiolated mesoporous silica compared to the aminated ones [23, 25].

Sol-gel process using polyethylene oxide (PEO) as well as low molecular weight surfactants have been adapted for the preparation of thermally stable, ordered and large pore size mesoporous non-silica materials of metal oxides including MnO_2, TiO_2, ZrO_2, Nb_2O_5, Ta_2O_5, Al_2O_3, SnO_2, WO_3, HfO_2, and mixed oxides of Al_2TiO_y, $ZrTiO_y$, ZrW_2O_y [26–28]. The pore topology and pore size of the mesoporous oxides can be controlled by varying the surfactant or by the addition of hydrophobic swelling agents to the template mixture. However, the nature of the inorganic precursors of transition metals plays a critical for this process and it has to be compatible with organic co-assembly [29]. Therefore the evaporation-induced self-assembly (EISA) was introduced for the synthesis of non-silica mesostructured materials and has shown to be the very useful approach for both controlling macroscopic form (powder, thin films, and membranes) and mesostructured [30, 31]. In this EISA process, the mixture of volatile solvent, (commonly ethanol) surfactant and inorganic precursors are exposed to evaporation which drives the co-assembly of surfactant molecules and inorganic precursors to form the ordered mesophases. Subsequently using aging, chemical or heat treatments induce the formation of the desired mesostructures. Recently mesostructures of titanium tungstate with the high surface area, large pore volume, uniform pore size, and tunable W/Ti ratios were successfully fabricated by using (EISA) strategy [32]. Also, mesoporous zirconium phosphotungstate with large specific surface area ~170 m^2/g, large pore volume, uniform pore size distribution, and tunable W/Zr ratios were prepared by adopting this EISA method [33]. In another template strategy, mesoporous TiO_2/WO_3 hollow fibers and interconnected nanotubes are prepared by hollow fibers as a framework, in addition, to block copolymer as a template. The mesostructures showed a surface area of ~130 m^2/g with improved photodegradation activity [34].

Earlier studies on the synthesis of inorganic ion-exchangers are well documented by Amphlet [35] and Clearfield [36]. A comprehensive review of inorganic ion-exchangers with layered structures [37] has been published which suggests that group IV phosphates, molybdates, tungstates, etc. possess interesting cation-exchange properties. Silicates of Ti(IV) have been shown to have porous properties and are being used as molecular sieves [38]. Two new crystalline phases of α-titanium phosphates with layered structures have been reported [39]. Siddiqui *et al.* [40] have synthesized crystalline phases of titanium tungstosilicate (TiWSi) and titanium tungstophosphate (TiWP) with better ion exchange capacity and better chemical and thermal properties as compared to titanium phosphate, titanium tungstate, and titanium silicate. Furthermore, ion-exchange materials with the amorphous state, crystalline state, and hybrid organic-inorganic states have been recently synthesized and studied for environmental applications [41, 42].

2. Experimental

2.1 Chemicals and materials

Triblock poly(ethylene oxide)-b-poly(propylene oxide)-b-poly-(ethylene oxide) copolymer Pluronic® P123 ($EO_{20}PO_{70}EO_{20}$, Mw = 5800, Sigma-Aldrich) was used as the structure-directing agent (SDA) and titanium tetraisobutoxide ($Ti(OBu)_4$, 98%, Aldrich), tungsten hexachloride WCl_6, phosphorus pentachloride PCl_5 (reagent grade, 95%, Sigma-Aldrich) and tetraethyl orthosilicate (TEOS, 98%, AnalaR) were used as Ti, W, P and Si source respectively. *n*-Propanol (n-C_2H_8O, BDH AnalaR) was used for dissolve the surfactant and hydrochloric acid HCl (37 wt. %, Analar) were all supplied by Shanghai Chemical Corp. Copper(II) nitrate hydrate, lead(II) nitrate (ACS reagent, ≥99.0%) and mercury(II) nitrate monohydrate (ACS reagent, ≥98.0%) are purchased from Sigma-Aldrich. All other chemicals were used as received without further purification and all aqueous solutions were prepared using Milli-Q (Millipore, Inc.) high quality deionized (DI) water (resistance 18.2 MΩ cm).

2.2 One-pot synthesis of bi-components mesoporous titanium tungstate (meso-TiW-x) using EISA approach

Mesoporous bi-components titanium tungstate (*meso*-TiW-x) oxides were synthesized with different composition through a modified evaporation-induced self-assembly (EISA) method. Typically, with a total Ti-W molar content of 18 mmol, $Ti(OBu)_4$ and WCl_6 were first dissolved in a mixture containing 20 g of *n*-propanol and 2.0 g of Pluronic 123 ($EO_{20}PO_{70}EO_{20}$). Then, the mixture was allowed for stirring for 2.0 hours at room temperature in a sealed vial. The obtained sol was cast into Petri dishes and kept under ambient conditions for 1 hour as shown in **Scheme 1**. Subsequently, the as-made gel was cured at 40°C for 2 days and then calcined in flowing air at 400°C for 4 hours with a ramping rate of 1°C min^{-1}. The obtained samples of *meso*-TiW were denoted as *meso*-TiW-x, where X equals 20, 30, and 40 wt. % the percentage of W in the original template mixture.

Specifically, for the synthesis of *meso*-TiW-20, *meso*-TiW-30 and *meso*-TiW-40, a 4.90, 1.43, 4.28 g of $Ti(OBu)_4$ were added to 2.14, 3.68, 2.86 g of WCl_6, respectively. In order to further moderate the hydrolysis rate, a 1.20 g of hydrochloric acid (37 wt. %) was added to the solution for synthesizing *meso*-TiW-20 oxides. A control experiment was performed to prepared pure *meso*-TiO_2 using EISA approach for comparison purpose.

2.3 Synthesis of mesoporous titanium tungstosilicate (meso-TiWSi) and titanium tungstophosphate mixed oxides (meso-TiWP)

The mesoporous titanium tungstosilicate oxides (*meso*-TiWSi-x, where x is equal 5, 10, 15, 20, 40, 60 wt. % the percentage of Si in the original template mixture) and titanium tungstophosphate mixed oxides (*meso*-TiWP-x, where x is the wt. % of P in template mixture equal 5, 10, 15, 20 and 30 wt. %) were synthesized through

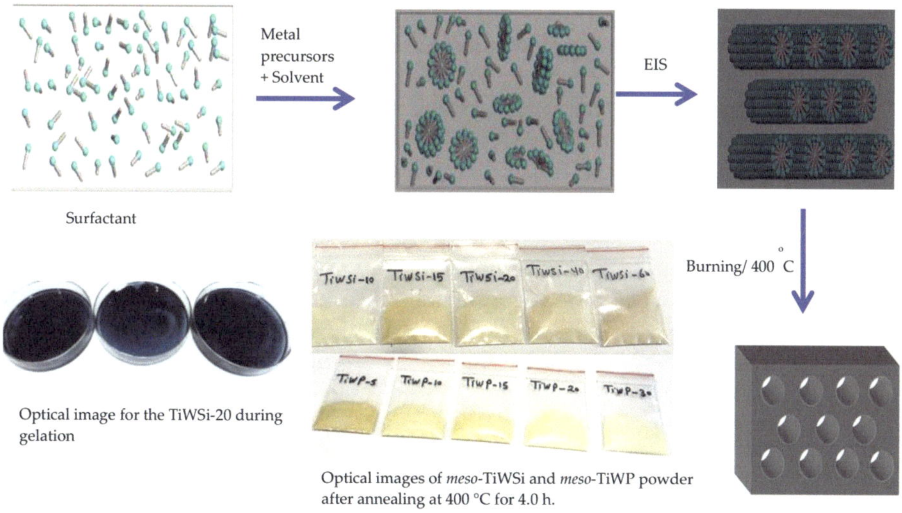

Scheme 1.
Schematic diagram for the synthesis of mesoporous materials of titanium tungstate (meso-TiW) titanium tungestophosphate (meso-TiWP) and titanium tungstosilicate (meso-TiWSi) by EISA.

adopted (EISA) method as shown above in **Scheme 1**. In a typical experiment, 2.86 g of WCl_6 was dissolved in 10 g of *n*-propanol, this mixture was mixed by stirring until getting the clear solution then 3.68 g of $Ti(OBu)_4$ was added to this solution to obtain a solution (A). Then 2.0 g of nonionic surfactant (triblock copolymer templates ($P^{®}123$), was dissolved in 10 g *n*-propanol and stirred for 10 min and different amounts of TEOS (5, 10, 15, 20, 40, 60 wt. %) were added to this solution and stirring for another 5.0 min to obtain a solution (B). In case of titanium tungstophosphate 5, 10, 15, 20 and 30 wt. % of PCl_5 was added to this solution and stirring for 5 min to obtain solution B. The solution (A) was added dropwise to (B) then the mixture was stirred for 2.0 hours at room temperature in a sealed vial. In order to further moderate the hydrolysis rate, 1.20 g of hydrochloric acid (37 wt%) was added to the solution for all prepared mixtures of mesoporous *meso*-TiWSi oxides. The obtained dark-blue colored homogeneous solution (shown in the inset in **Scheme 1**) was cast into Petri dishes and kept under ambient condition for 1.0 hour. Subsequently, the as-made gel was cured at 40°C for 2 days to completely remove solvents. Finally, the blue deposit was scraped and crushed into powder and then calcined in flowing air at 400°C for 4.0 hours with a ramping rate of 1.0°C/min.

2.4 Materials characterizations

X-ray powder diffraction (XRD) and Small-angle X-ray (SAXS) measurements were recorded respectively on a Bruker D4 powder X-ray diffractometer (Germany) and Nanostar U small-angle scattering system (Bruker, Germany) by using Cu K_{α} radiation (40 kV, 40 mA). The BET surface area, pore volume, and the pore size distributions were measured by nitrogen adsorption/desorption using a Micromeritics Tristar 3020 analyzer (USA). Prior to the measurements, the samples were degassed at 180°C for at least 6.0 hours. The Brunauer-Emmett-Teller (BET) method was utilized to calculate the specific surface area. By using the Barrett-Joyner-Halenda (BJH) model, the pore volumes and pore size distributions were derived from the adsorption branches of isotherms. Transmission electron microscopy (TEM) analysis was conducted on a JEOL 2011 microscope (Japan) operated at 200 kV and equipped with an energy dispersive X-ray (EDX) detector. Field-emission scanning electron microscopy (FESEM) images were collected on the Hitachi Model S-4800 field-emission scanning electron microscope. The heavy metals analysis was performed using an inductively coupled plasma optical emission spectroscopy (ICP-OES) system (Thermo Scientific; UK, Model ICAP 6000).

2.5 Procedure for heavy metal ions adsorption from water by mesoporous materials

The standard procedure for the removal of heavy metal ions from water using the produced mesoporous materials was performed by mixing a weighted amount of the mesoporous absorbent with 3.0 mL solutions containing 0.2 mg/L Cu(II), Cd(II) or Pb(II) nitrate salt in polypropylene tubes. The samples were placed in a shaking water bath at 25°C for 2.0 hours and then the solution was separated by centrifuge at rpm 7000 for 10 min in each run. Each solution was then diluted by taking 100 μl and transferred into 10 mL conical polypropylene flask for analysis by ICP. The amount of metal ion adsorbed at the mesoporous adsorbents was calculated by the difference of the initial concentration (C_o) and the equilibrium concentration (C_e). The removal efficiency and equilibrium adsorption capacity of Cd(II), Pb(II), and Hg(II) ions from the solution was calculated using the following equations:

$$(E\%) = (C_o - C_e/C_o) \times 100 \qquad (1)$$

$$(Q_e\%) = (C_o - C_e)\,V/m \qquad (2)$$

Where, E, Q_e, C_o, C_e, V, and *m* are the removal efficiency, equilibrium adsorption capacity (mg/g), initial concentration (mg/L), equilibrium concentration (mg/L), solution volume (L) and adsorbent mass (g) respectively. To investigate metal ion loss during the adsorption due to the glassware or any of the experimental equipment a parallel experiment was performed in the absence of the mesoporous adsorbents and it was regarded as a standard solution.

3. Results and discussion

3.1 Synthesis and characterizations of bi-components meso-TiW prepared by EISA approach

In this section, we present the result for the synthesis of bi-components *meso*-TiW with various ratios using modified solvent evaporation-induced self-assembly (EISA). Firstly the synthesis of bi-components of mesoporous titanium tungstate (*meso*-TiW) with various compositions was executed. In EISA process the mixture of volatile solvent, (commonly ethanol), copolymer surfactant and inorganic precursors are mixed together and exposed to evaporation which drives the co-assembly of surfactant molecules and inorganic precursors to form the ordered mesostructures. The W content was varied between 20 and 40 moles and the *meso*-TiW powder after annealed at 400°C for 4.0 hours shows the light yellow color crystal as such shown in the inset of **Scheme 1** which indicating on the precipitation of TiW oxide while pure *meso*-TiO_2 exhibits white color.

Figure 1 shows the wide-angle XRD patterns for *meso*-TiW-20 and for pure *meso*-TiO_2 prepared using EISA approach for comparison purpose. The XRD pattern of pure *meso*-TiO_2 displayed a well-crystallized anatase phase (JCPDS Card No. 01-083-2243), while for *meso*-TiW-20 the XRD peaks intensity is slightly decreased in addition to the disappearance of some other peaks. This can attribute to the addition of 20 wt. % of W precursor hinders the crystallization of anatase phase and instead an amorphous composite of Ti—W oxides were formed.

The nitrogen adsorption-desorption isotherms and the plot for pore size distribution of the mesoporous TiW-x oxides are shown in **Figure 2**. From the isotherms we can see that the *meso*-TiW oxide showed typical type IV isotherms with sharp capillary condensation steps at a relative pressure (P/P^o) of 0.35–0.85 that characteristic for mesoporous materials.

The average pore size, surface area (BET), pore volume for the *meso*-TiW oxide sample in comparing with pure mesoporous TiO_2 (prepared using EISA approach but without adding WCl_6) are shown in **Table 1**. Obviously, the *meso*-TiW-30 oxide shows the highest surface area; lower pore volume and a narrow pore size distribution which demonstrates the formation of highly ordered and uniform mesoporous structure. On the other hand, the pure mesoporous TiO_2 prepared in absence of WCl_6 shows lower surface area (81 m^2/g) and higher pore volume with bimodal pore diameters. This can be attributed to the incorporation of W precursor effectively improve the order of mesoporous structure and because of the tungsten oxide has a high density (7.16 g/cm^3) the incorporation of W leads to slightly reduce the pore volume of *meso*-TiW oxide.

Figure 3 shows the TEM image of the mesoporous *meso*-TiW samples. Clearly, the fine structure of the *meso*-TiW oxides shows typical morphologies of the highly ordered 2D hexagonal mesostructures with highly orderly channels extended over large domains which are consistent with high surface area obtained by N_2 adsorption results and confirms the successful production of bi-components mesoporous TiW oxides by EISA approach.

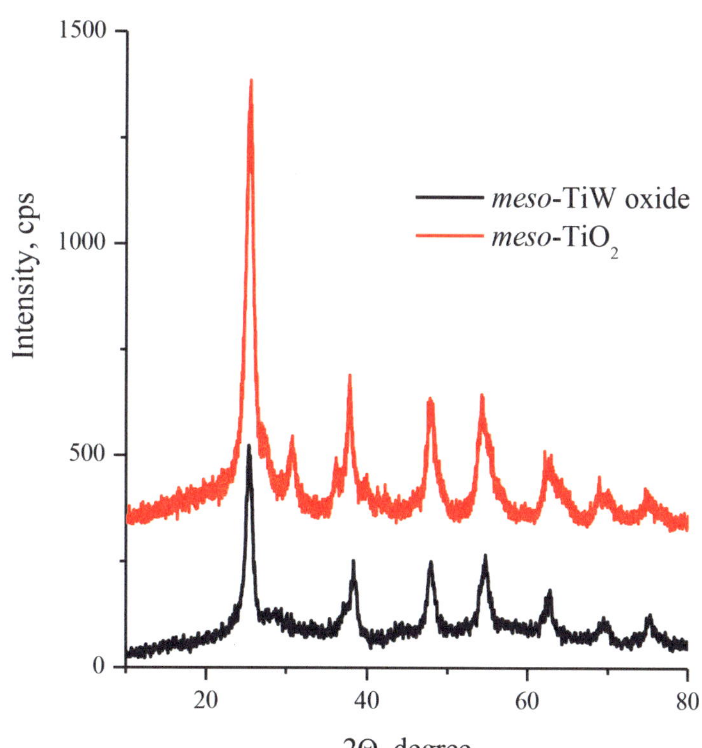

Figure 1.
X-ray diffraction patterns for meso-TiW-20 and pure meso-TiO_2 prepared by EISA approach and annealed at 400°C, 4.0 hours.

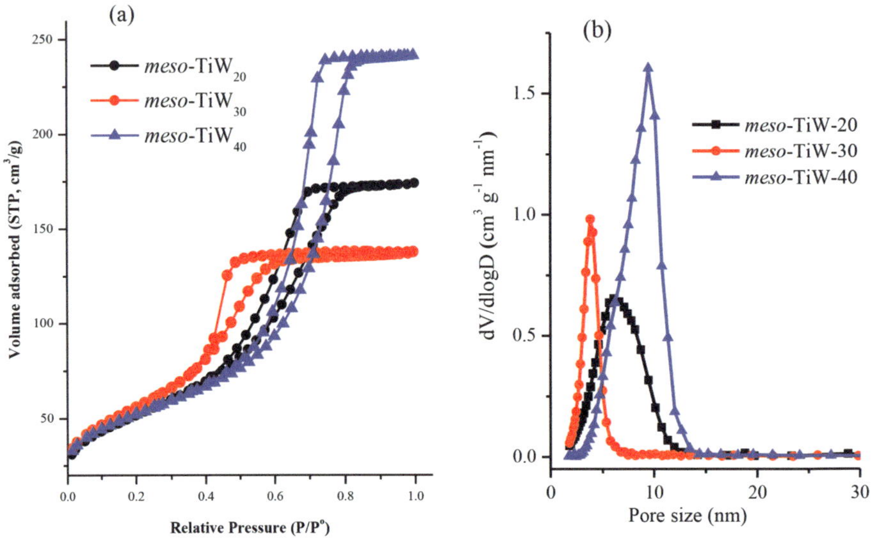

Figure 2.
(a) Nitrogen adsorption-desorption isotherms, (b) the corresponding pore size distribution curves of the meso-TiW oxide.

Samples	SBET (m^2/g)	Pore volume (cm^3/g)	Pore size (nm)
meso-TiW-20	189	0.27	6.1
meso-TiW-30	206	0.21	3.9
meso-TiW-40	188	0.37	9.5
meso-TiO_2	81	0.33	6/12.5

Table 1.
The textural properties of the samples of mesoporous TiW oxides.

3.2 Synthesis and characterization of tri-component mesoporous titanium tungstosilicate (meso-TiWSi) materials

Here we investigate the possibility of synthesis more complicated mesoporous materials of tri-components mesoporous oxides of Ti, W, and Si. Because the bi-components *meso*-TiW-30 showed the highest surface area (206 m^2/g) and ordered pore structure we used its composition of mole ratio ($Ti_{70}W_{30}$) as start material for the synthesis of tri-component mesoporous titanium tungstosilicate *meso*-TiWSi-x and tungstophosphate (*meso*-TiWP-x) materials with various compositions of silica (x = 10, 15, 20, 40 and 60 wt. %) via the facile EISA method employing P®123 surfactant, $Ti(OBu)_4$, WCl_6 and TEOS as described in Section 2 above and shown in **Scheme 1**. Optical images of the produced *meso*-TiWSi materials with Si ratio of 10, 15, 20, 40 and 60 wt. % after annealing at 400°C in the air are shown in the inset of **Scheme 1**.

Figure 4 shows the low-angle and wide-angle X-ray diffraction patterns of mesoporous *meso*-TiWSi-x calcined at 400°C in the air. As shown in **Figure 4a**, there's a distinct peak at $2\theta = 0.53°$can be observed which is indexed as (100) reflections characterize of the 2D hexagonal mesostructure [32, 43]. Furthermore, another one broad and weak diffraction peak can clearly observe at $2\theta = 1–1.5°$ indicates the long-range order of the mesopores structure. An increase in the Si ratio up to 60 wt. % result in one broad peak can be seen in the SXRD diffraction pattern, suggested that an excessive increase in Si incorporation decreased the ordering of the mesoporous structure. For the TiWSi-10, the first scattering peak shifts from a 0.53 nm^{-1} to

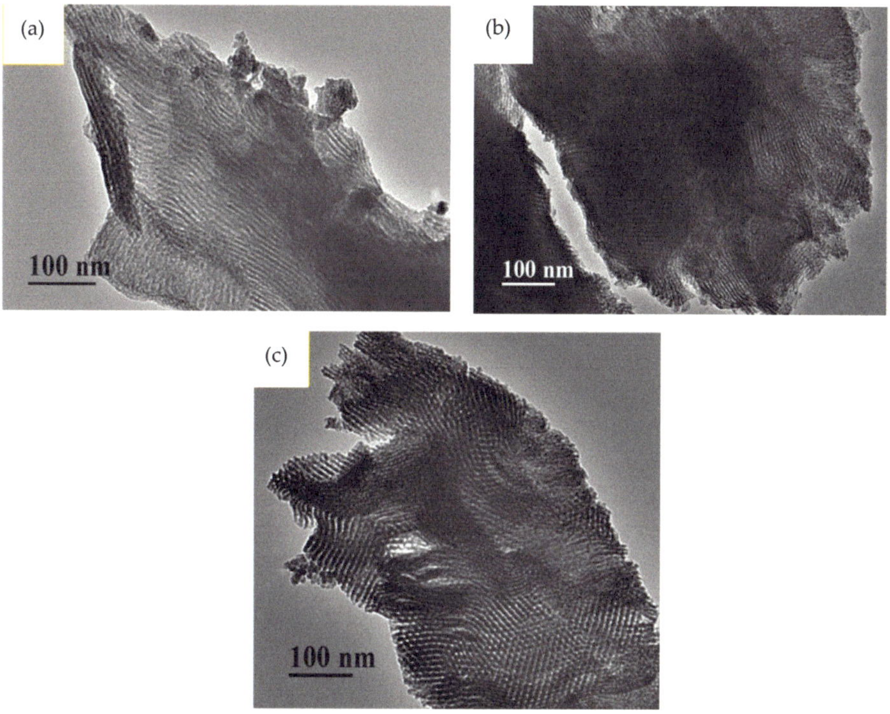

Figure 3.
TEM images of the samples (a) meso-TW-20, (b) meso-TW-30, and (c) meso-TW-40.

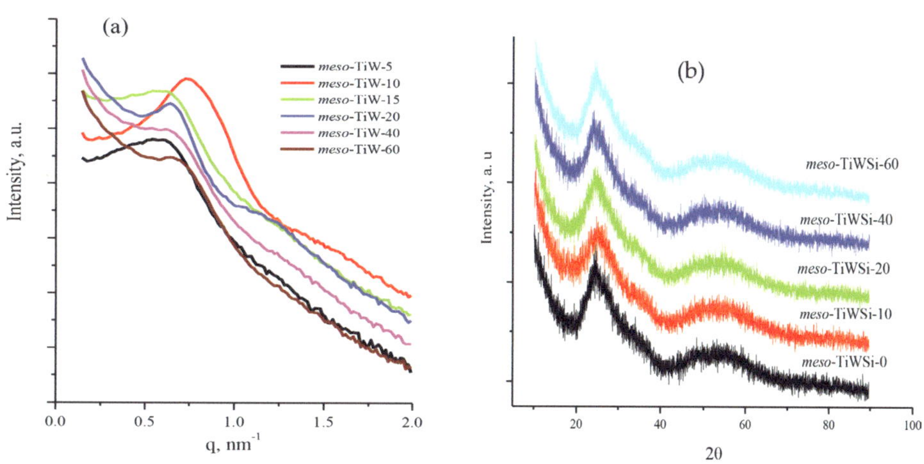

Figure 4.
(a) Low-angle and (b) wide-angle X-ray diffraction patterns of mesoporous meso-TiWSi-x calcined at 400°C in the air for 4.0 hours.

a larger *q* value of a 0.56 nm^{-1}, corresponding to the framework shrinkage during the removal of the template. The wide angle X-ray diffraction (WXRD) patterns of these samples are shown in **Figure 4b**. All the patterns kept consistent with each other and only exhibited two broad peaks in the 2θ range of 20–40° and 40–60°, indicating the existence of an amorphous or semicrystalline mesoporous TiWSi wall structure. No individual crystalline peak of titanium tungstate or titanium was observed.

Meanwhile, there were no peaks of tungsten oxide in the patterns, illustrating the high dispersion of tungsten and silicon oxide among the mesoporous framework even in *meso*-TiWSi-0, which had silica ratio of 0 wt. %. This might be accounted to the silica atoms in the materials were segregated by titanium and tungsten atoms due to the advantage of using the one-pot EISA method [44]. We presumed that the presence of W and Si species somehow retarded the crystallization of TiO_2 during the calcination and hence benefited the preservation of the ordered mesostructured which is consistent with **Figure 4b** and the previous studies [45, 46].

In order to further confirm the presence of mesopores in *meso*-TiWSi-x materials, the SEM and TEM characterizations were performed and **Figure 5** shows the SEM images of the *meso*-TiWSi-x samples with different Si ratios after the annealing process. For the *meso*-TiWSi-10 (**Figure 5a**) and *meso*-TiWSi-20 (**Figure 5b**), the SEM images clearly show the highly ordered mesopores uniformly exist through the TiWSi framework over a wide micrometer area. However for *meso*-TiWSi-40 and *meso*-TiWSi-60, the SEM images in **Figure 5c** and **d** show wormlike mesoporous structure indicating on less order pore arrangement when Si ratio is increased to 60 wt. %.

Figure 6 shows the TEM images for different composition of *meso*-TiWSi after calcination and obviously, the TiWSi framework exhibits a 2D hexagonal mesostructured with ordered mesopores of about 10 nm in diameter. It also clearly observes the existence of ordered pores along the [110] and [001] directions in **Figure 6b** of *meso*-TiWSi-20. Nevertheless by further increasing the Si ratios to 60 wt. % (**Figure 6d**), the structure showed wormlike mesoporous channels rather than the long-range ordered hexagonal pores. The results of SEM and TEM agreed quite well with the results of the SXRD characterization, which predicted the presence of ordered mesoporous structure for samples with Si ratio from 0 to 20% and wormlike mesopores structure for meso-TiWSi-60.

For the elemental composition analysis, the backscattering SEM image and energy-dispersive X-ray (EDX) analysis of the elemental composition of the *meso*-TiWSi-20 materials are shown in **Figure 7**. The backscattering SEM image (**Figure 7a**) obviously shows very and highly order channels running through the *meso*-TiWSi framework confirming the formation of long-range and ordered

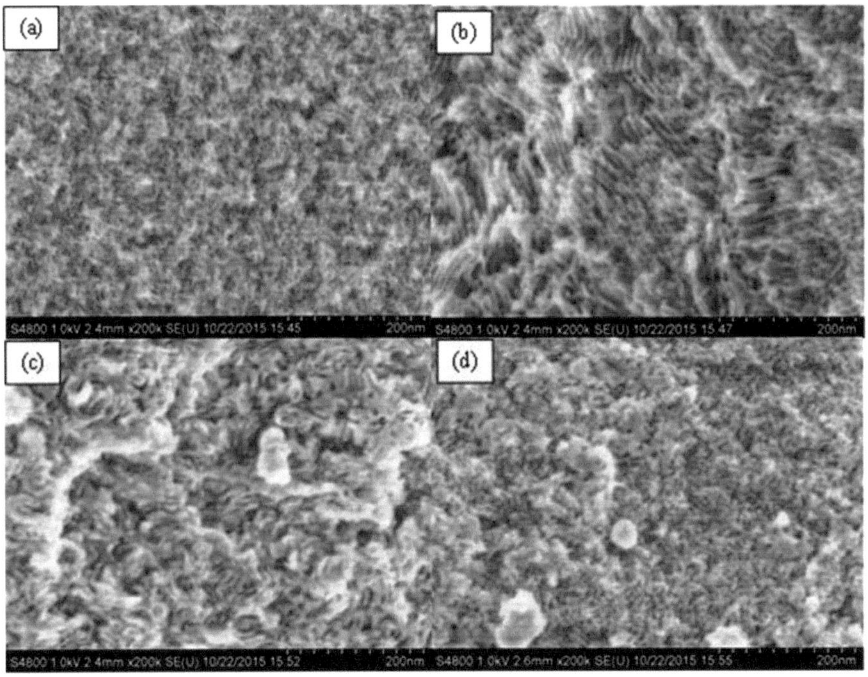

Figure 5.
SEM images of (a) meso-TiWSi-10 (b) meso-TiWSi-20 (c) meso-TiWSi-40 (d) meso-TiWSi-60.

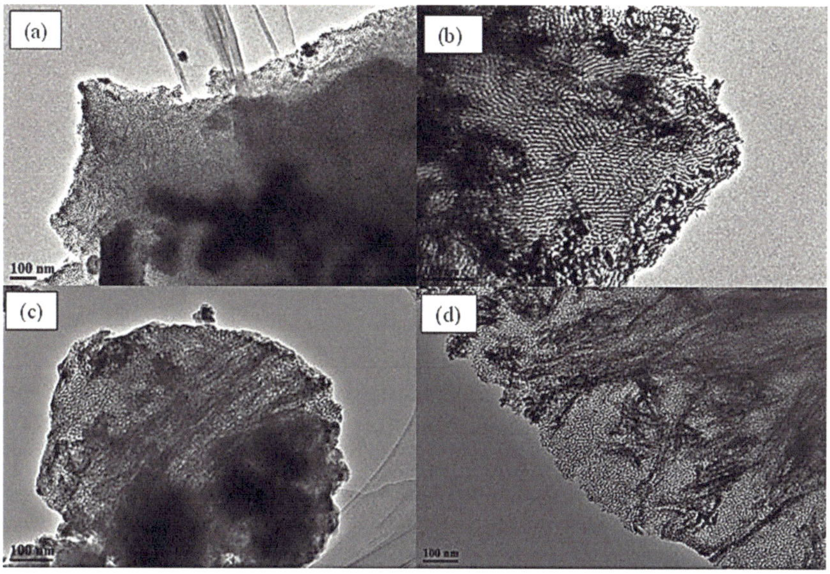

Figure 6.
TEM images of (a) meso-TiWSi-10 (b) meso-TiWSi-20 (c) meso-TiWSi-40 (d) meso-TiWSi-60.

mesopores. The EDX elemental analysis in **Figure 7b** shows evidently the existence of Ti, W, O and Si elements approving that all the elements were successfully introduced into the framework of the mesoporous structure. The Si wt. % derived from the EDX analysis were very close to the target composition set up in the template mixture as shown in **Table 2**.

The textural properties and the surface area of the ordered *meso*-TiWSi-x with different Si ratios were characterized by the N_2-physisorption technique. Nitrogen adsorption and desorption isotherms and the corresponding pore size distributions of the *meso*-TiWSi materials are shown in **Figure 8**. For all the materials, the isotherms exhibited typical type IV isotherms and well-defined H_2 shaped hysteresis loops with sharp capillary condensation steps at a relative pressure (P/P^o) of 0.4–0.8, that characteristic of mesoporous materials [45, 47]. Besides, all the capillary condensation steps of the hysteresis loops for all the samples were very steep, pointing to the presence of a uniform and ordered pore size in the framework of *meso*-TiWSi. The average pore size of these samples was narrowly distributed at around 4–10 nm, implying the existence of the uniform mesoporous channels within the obtained materials. The details textual parameters including specific surface area, pore size and volume of *meso*-TiWSi-x obtained from the N_2 adsorption-desorption isotherms, are summarized in **Table 3**. Clearly, all the samples

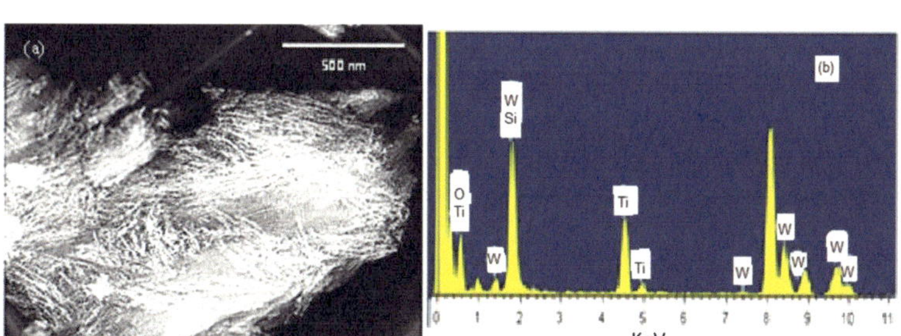

Figure 7.
(a) Backscatter SEM image of meso-TiWSi-20 (b) EDX elemental profile of meso-TiWSi-20.

Samples	Si wt. %
meso-TiWSi-10	7.4
meso-TiWSi-20	18.4
meso-TiWSi-60	54.3

Table 2.
EDX analysis of the Si wt. % in the meso-TiWSi samples.

had excellent textural properties with large specific surface area (131–188 $m^2 g^{-1}$), uniform pore size (4–10 nm) and pore volume (0.17–0.37 $cm^3 g^{-1}$).

3.3 Synthesis and characterization of tri-component mesoporous titanium tungstophosphates (meso-TiWP-x) materials

Tri-components mesoporous titanium tungstophosphates *meso*-TiWP-x materials were fabricated via the facile EISA method employing P®123 surfactants as a template and $Ti(OBu)_4$, WCl_6 and PCl_5 as precursor for Ti, W and P respectively. Starting with Ti: W mole ration of 70:30, various compositions of P (5, 10, 15, 20 and 30 wt. %) were incorporated into the *meso*-TiW-30 and an optical image of the produced *meso*-TiWP-x samples is shown in the inset of **Scheme 1**.

Figure 9 shows the small and wide angle X-ray diffraction pattern of *meso*-TiWP-x samples after annealing at 400°C for 4 hours. As indicates in **Figure 9a** there's a broad scattering peak in the 2θ range 0.5–1° which indicates on the formation of less ordered mesoporous TiWP structures. However, the *meso*-TiWP-10 shows rather well-defined peak among the other samples; presumably, due to better order pores have been formed. The wide-angle X-ray diffraction patterns of the TiWP samples shown in **Figure 9b** are consistent with each other and only exhibited two broad peaks in the 2θ range of 20–30° and 40–60° which is evidence for the existence of an amorphous or semi-crystalline framework of mesoporous TiWP. Based on the above results it can be concluded that the produce *meso*-TiWP-x samples have amorphous and less order porous structures than in the case of TiWSi.

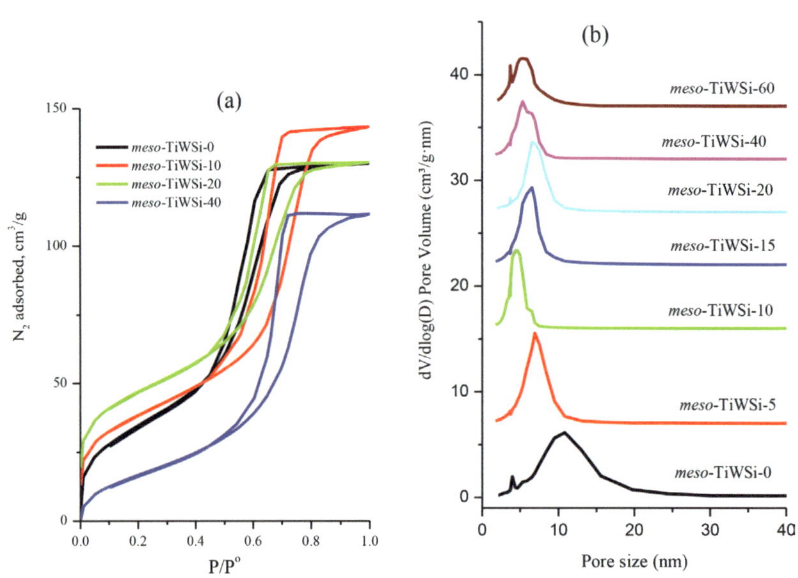

Figure 8.
(a) Nitrogen adsorption isotherms and (b) the corresponding pore size distribution curves of the meso-TWSi materials.

Samples	Specific surface area ± 5 ($m^2 g^{-1}$)	Pore size ±0.5 (nm)	Pore volume ($cm^3 g^{-1}$)
meso-TiWSi-0	188	9.5	0.37
meso-TiWSi-5	131	7.0	0.20
meso-TiWSi-10	164	5.5	0.19
meso-TiWSi-15	138	7.0	0.20
meso-TiWSi-20	142	7.5	0.22
meso-TiWSi-40	151	6.0	0.17
meso-TiWSi-60	156	5.0	0.19

Table 3.
Textural properties of the meso-TWSi-x materials derived from nitrogen adsorption and desorption analysis.

Figure 10 reveals the surface morphology of the *meso*-TiWP-x samples and by examining the SEM images, the formation of the pores are evidently observed within the TiWP framework and generally, the pores order seems to be irregular. However, for the *meso*-TiWP-10 sample, it shows some degree of ordered pores domains is formed. In addition the X-ray mapping and EDX composition analysis (not shown here) demonstrate the presence of Ti, W, P and oxygen elements which are uniformly distributed through the mesoporous TiWP materials.

The transmission electron microscope characterization of the *meso*-TiWP-x materials is illustrated in **Figure 11**. The TEM images of the mesoporous TiWP-x materials display large domains of the porous structure obtained after direct calcination at 400°C in the air due to the removal of templates. Clearly, the pores arrangement is rather random presumably due to the source of phosphor (PCl_5) affected the formation of the ordered mesostructured during the self-assembly process. However the TEM image for *meso*-TiWP-10 with phosphor content of 10 wt. % shows the pores are arranged in the highly ordered structure. It implies that adding phosphorus with 10 wt. % retains the pore order of the titanium tungstophosphates.

The N_2 adsorption/desorption isotherms of *meso*-TiWP-x materials after annealing at 400°C in the air are shown in **Figure 12a**. Evidently, the N_2 adsorption/

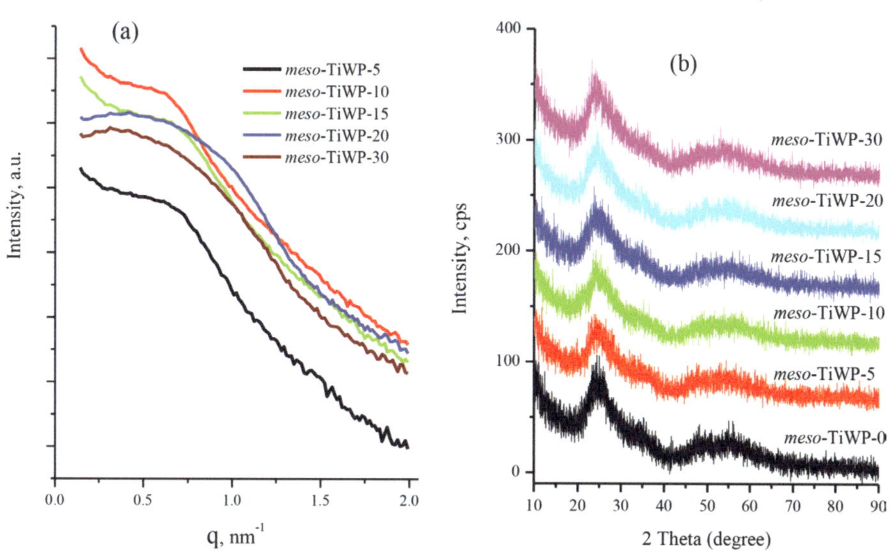

Figure 9.
X-ray diffraction patterns for meso-TiWP-x materials, (a) low angle and (b) high angle.

desorption isotherms for *meso*-TiWP-x materials show typical type IV curves with a sharp capillary condensation step at $P/P^o = 0.8–0.9$ and an H_2-type hysteresis loop, which is typical of large-pore mesoporous materials with cylindrical channels.

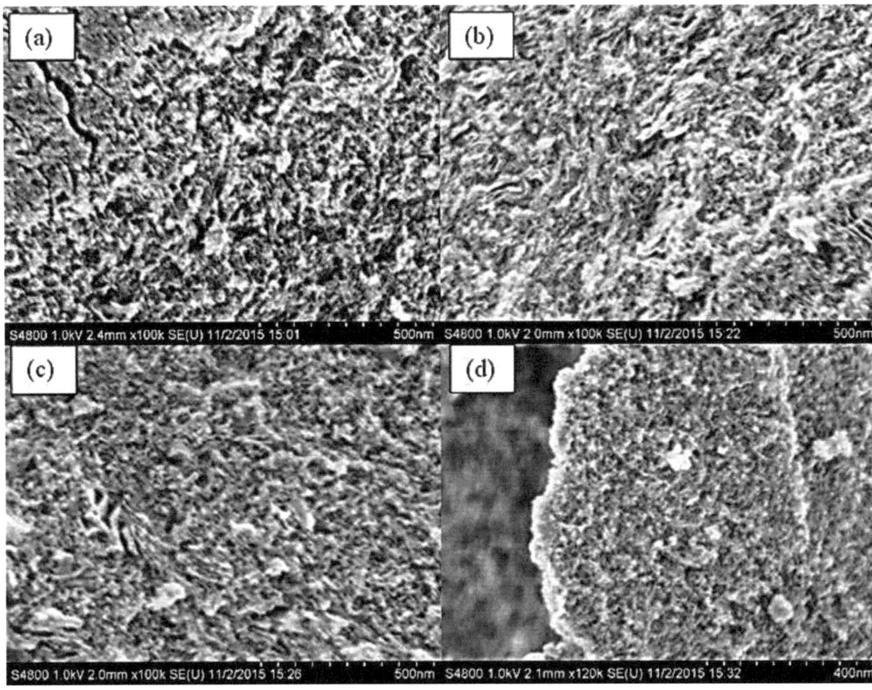

Figure 10.
SEM images of meso-TiWP-X: (a) meso-TiWP-5 (b) meso-TiWP-10 (c) meso-TiWP-15 (d) meso-TiWP-20.

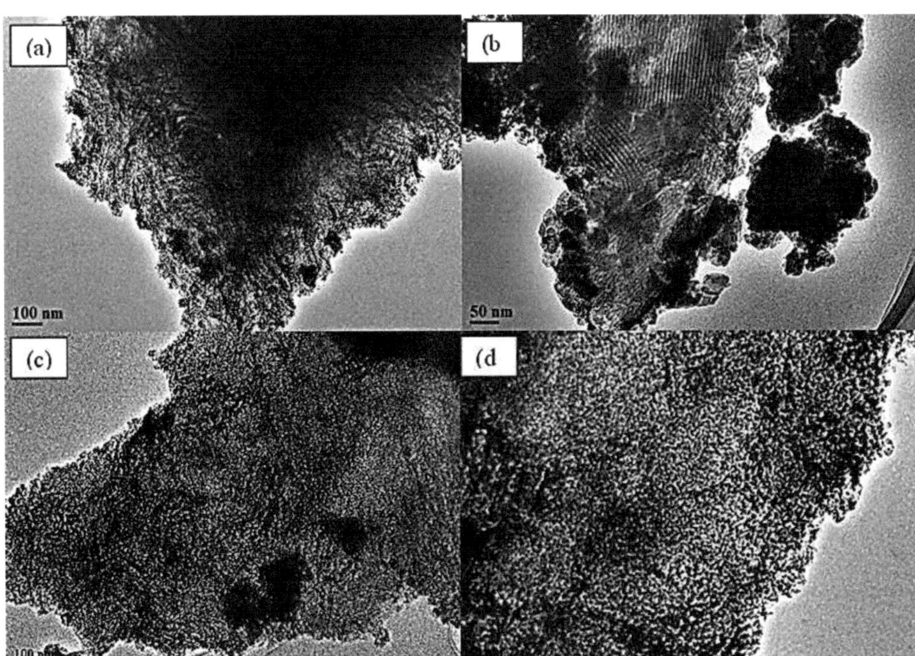

Figure 11.
TEM images of (a) meso-TiWP-5, (b) meso-TiWP-10, (c) meso-TiWP-15, (d) meso-TiWP-20 after annealing at 400°C in air for 4 hours.

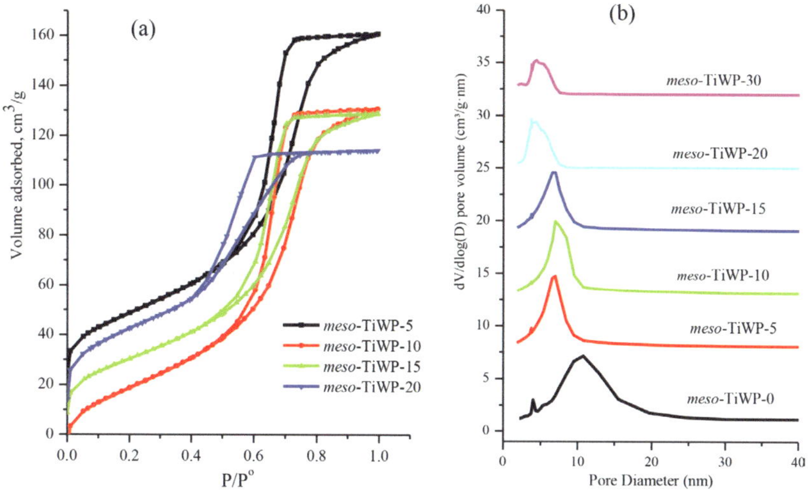

Figure 12.
(a) Nitrogen adsorption/desorption isotherms of meso-TiWP-x and (b) the corresponding pore size distribution curves.

The pore size derived from the adsorption branches of the isotherms by using the Barrett-Joyner-Halenda method is shown in **Figure 12b**. Generally, the pore diameter for the *meso*-TiWP-x materials shows a narrow distribution and a gradual decrease as the phosphorus content increases, with a pore diameter of 9.6 nm for *meso*-TiWP-0 and 4.3 nm for *meso*-TiWP-30 as reported in **Table 4**. The Brunauer-Emmett-Teller surface area and pore volume of the *meso*-TiWP-x materials are calculated from the N_2 adsorption/desorption isotherms and reported in **Table 4**. The BET surface area of the meso-TiWP-x materials is varied between 188 m^2/g for no phosphorus and a 128 m^2/g for *meso*-TiWP-15 which presumably due to the formation of the more crystalline mesoporous wall as the phosphorus is added to titanium tungstate. However, it can conclude that mesostructured of titanium tungstate can be retained during the addition of phosphorus and annealing treatment which confirms the versatility of EISA soft template approach to producing multi-components mesoporous oxides with various compositions.

3.4 Removal of heavy metal ions from water by mesoporous titanium tungstosilicate and tungstophosphate materials

This section reports the preliminary evaluation of the produced mesoporous materials for heavy metal ions removal from synthetic water solution. The ability

Material	Pore diameter/ nm	Pore volume/ $cm^3 g^{-1}$	Specific surface area/ $m^2 g^{-1}$
meso-TiWP-0	9.5	0.37	188
meso-TiWP-5	7.5	0.232	140
meso-TiWP-10	7.6	0.220	129
meso-TiWP-15	5.6	0.200	128
meso-TiWP-20	4.5	0.150	153
meso-TiWP-30	4.3	0.120	154

Table 4.
Textural properties of the meso-TiWP-x materials derived from nitrogen adsorption/desorption analysis.

of mesoporous titanium tungstosilicate to enrich the metals ions absorption from water was tested by adding aliquots of metal ions solution to a known weight of *meso*-TiWSi-x absorbent. **Figure 13** shows the effect of the *meso*-TiWSi-x loading on the removal efficiency of Cd(II), Hg(II) and Pb(II) ions at pH equals 7 and 25°C. It is observed that the Cd(II) ion uptakes are independent on absorbent loading of *meso*-TiWSi-10 (**Figure 13a**) and the removal efficiency stabilized at around 92%. On the other hand, for Hg(II) ion uptakes gradually increase as the meso-TiWSi-10 absorbent load increases reaching about 92% at 200 mg *meso*-TiWSi-10 loading. However, for Pb(II) ion adsorption at *meso*-TiWSi-10 is inconsistent and unexpectedly shows a decrease in absorption efficiency as the absorbent loading increased. **Figure 13b** shows with the addition of *meso*-TiWSi-20 adsorbent in the range of 50–200 mg, no significant variation in the metal ions uptakes were observed for Hg(II), whereas slight decrease in Cd(II) ions removals was maintained. In contrast, the Pb(II) ions uptakes show a remarkable increase in a number of *meso*-TiWSi-20 adsorbent increases and 80% removal efficiency was achieved at 200 mg

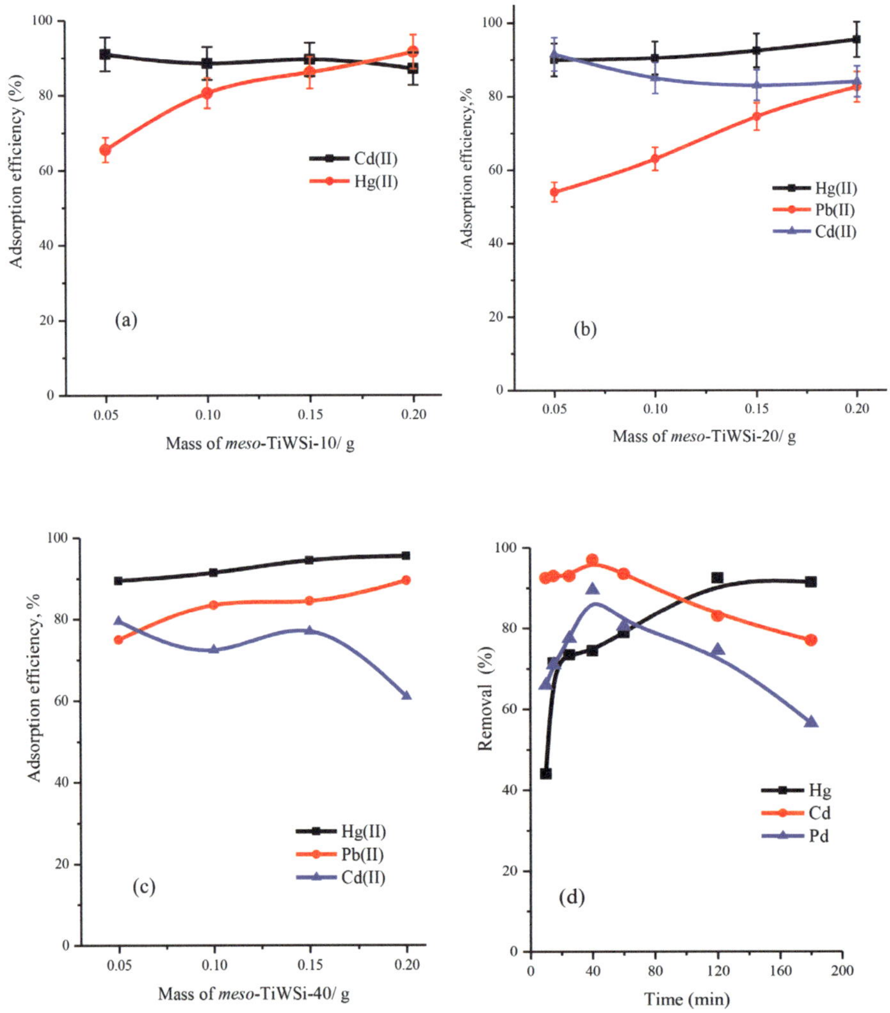

Figure 13.
Effect of the adsorbent dose of (a) meso-TiWSi-10, (b) meso-TiWSi-20, (c) meso-TiWSi-40 on the adsorption of Pb(II), Cd(II) and Hg(II) from aqueous solution at pH of 7, a contact time of 2 hours and a metal ion concentration of 20 ppm, (d) effect of contact time on the adsorption of Pb(II), Cd(II) and Hg(II) from aqueous solution by m-TiWSi-20 at pH of 5, 0.15 g m-TiWSi-20/3 mL solution and a metal ion concentration of 0.2 mg/L.

adsorbent. Effect of *meso*-TiWSi-40 adsorbent dose on the adsorption of Cd(II), Pb(II) and Cd(II) are shown in **Figure 13c**. The removal efficiency of Hg(II) and Pb(II) gradually increases with increasing the adsorbent dose and record about 90 and 85% respectively. This can attribute to the greater availability of the exchangeable sites or surface area at a higher concentration of the adsorbent. While increasing the *meso*-TiWSi-40 adsorbent dose gave rise to a remarkable decrease in adsorption of Cd(II).

Figure 13d shows the effects of agitation time on the removal efficiency of Hg(II), Cd(II) and Pb(II) using 150 mg of *meso*-TiWSi-20 adsorbent at pH = 6. It was noted that the adsorption of Cd(II) and Pb(II) gradually increases with an increase in agitation time and attains an equilibrium adsorption of 95 and 89% respectively within 40 min. Then the removal efficiency gradually decreases at longer agitation time. This result is important as the equilibrium time is a key parameter for economical wastewater treatment applications. On the other hand, the adsorption equilibrium time for Hg(II) takes about 2 hours to reach adsorption efficiency of 90%.

The mesoporous titanium tungstophosphate (*meso*-TiWP) adsorbent was also evaluated for the adsorption of heavy metal ions from water. **Figure 14a** shows the effect of the *meso*-TiWP-15 and *meso*-TiWP-20 adsorbent dose of on the uptakes of Hg(II) ions from water at pH 6. At a fixed Hg(II) ion concentration, it can be observed that the removal % of Hg(II) ion slightly increases with increasing weight of the adsorbents. This is attributed to the high surface area and greater availability of adsorption active sites of *meso*-TiWP-15 and *meso*-TiWP-20 adsorbents. For 95% removal of Hg(II) ion a maximum of 100 mg is required from *meso*-TiWP-15 or *meso*-TiWP-20 adsorbent. **Figure 14b** shows the effect of adsorption time on the removal efficiency of Hg(II), Cd(II) and Pb(II) ions using 150 mg of the *meso*-TiWP-15 absorbent. It is seen that for *meso*-TiWP-15 absorbent exhibited a rapid adsorption for Cd(II) and Pb(II) ions with a removal efficiency reaching about 90% within the first 10 min while it takes about 1.0 hour to achieve 80% for Hg(II) adsorption.

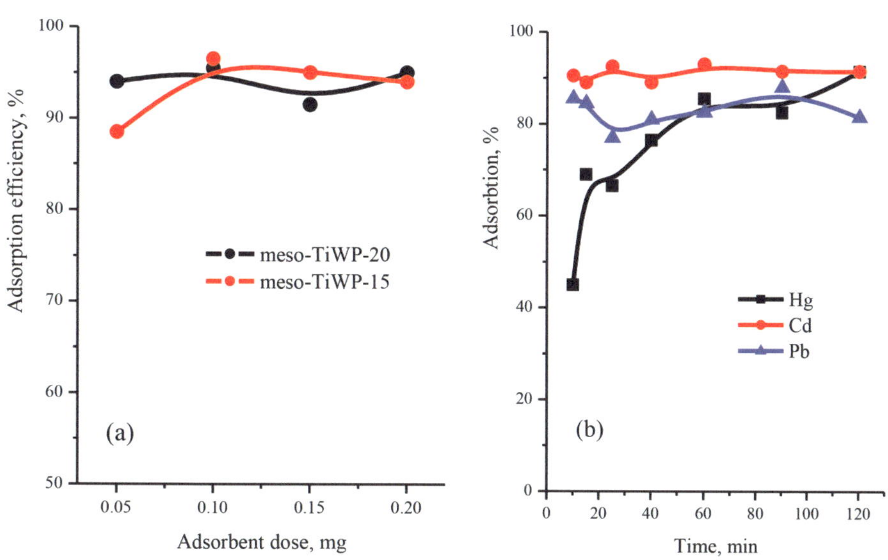

Figure 14.
(a) Effect of the meso-TiWP-15 and a meso-TiWP-20 dose of on the adsorption of Hg(II) from aqueous solution, experimental conditions included a solution pH of 6, a contact time of 2 hours and a metal ion concentration of 20 mg/L, (b) effect of contact time on the adsorption of Pb(II), Cd(II) and Hg(II) from aqueous solution by m-TiWP-15. Experimental conditions included a solution pH of 5, 0.15 g m-TiWP-15/3 mL solution and a metal ion concentration of 0.2 mg/L.

Adsorbent	Cu	Pb	Cd	Hg	Ref.
SBA-15-3-aminopropyl trimethoxy-silane	98.0	99.0	90.0	—	[48]
KIT-6-ethylenediaminetetraacetic acid (EDTA)	99.5	98.0	99.0	—	[49]
SBA-15-ethylenediaminetetraacetic acid (EDTA)	99.0	98	99.5	—	[49]
SBA-15-aminopropyl (SBA/NH_2)	—	96	50	—	[50]
Magnetic MCM-48 mesoporous silica amine (—NH_2)	91	96	90	—	[51]
meso-TiWP-20	95.9	95.5	95.8	—	This work
meso-TiW-20	50	76	—	—	This work
meso-TiWSi-10	—	82	91	92	This work

Table 5.
A comparison of the heavy metal ions removal efficiency of the best performance meso-TWSi-x and meso-TiWP-x materials with some of the modified mesoporous silica-based material reported in the literature.

Finally, **Table 5** shows a comparison of the heavy metal ions removal efficiency of the best performance of our mesoporous titanium tungstophosphate and tungstosilicate materials with some of the modified mesoporous silica-based material widely reported in the literature at similar experimental conditions. Mesoporous *meso*-TiWP-20 showed a good performance for removing Cu^{2+}, Pb^{2+}, and Cd^{2+} in the pH region of 5.0–7.0 which is comparable to the performance, for example, mesoporous silica modified with amine functional group and magnetic iron oxide. This metal ion adsorption enhancement can be apparently attributed to the higher surface area and active sites available to metal ion binding at the surface of *meso*-TiWP and *meso*-TiWSi adsorbents [7].

4. Conclusion

In this work we investigate the synthesis of new bi- and tri-component of mesoporous oxides of titanium tungstate (TiW), titanium tungstosilicate (TiWSi) and titanium tungstophosphate (TiWP) oxides. For the mesoporous titanium tungstosilicate, the silica content was varied between 10, 15, 20, 40 and 60 wt. % using TEOS precursor, while for mesoporous titanium tungstophosphates the ratio of phosphorus was varied as 5, 10, 15, 20 and 30 wt. % by adding PCl_5 into the EISA template mixture of P®123 surfactants, $Ti(OBu)_4$ and WCl_6 precursors. Subject to silica and phosphorus content, the obtained *meso*-TiWSi-x and *meso*-TiWP-x exhibits high surface area, ordered 2D hexagonal mesostructure with orderly channels extended over a large domain. The EISA approach demonstrates a successful synthesis approach for the synthesis of new complex inorganic nanostructured and mesoporous materials for potential application in environmental and catalysis. The produced *meso*-TiW, *meso*-TiWSi, and *meso*-TiWP adsorbents exhibit good adsorbent efficiency for the removal of Pb(II), Cd(II) and Hg(II) ions from water due to the presence of high surface area and accessibility of surface active sites. The adsorption of these metal ions is found to be dependent on contact time and adsorbents dose.

Acknowledgements

This work was funded by The National Plan for Science, Technology, and Innovation (MAARIFAH) King Abdulaziz City for Science and Technology, Kingdom of Saudi Arabia, Award number AT34-203. The authors appreciate Dr. Yuhui Li and Prof. Dongyuan Zhao from Fudan University for their cooperation in the SEM and TEM analysis.

Conflict of interest

There is no any conflict of interest can be declared.

Author details

Mohamed A. Ghanem*, Abdullah M. Al-Mayouf and Mabrook S. Amer
Electrochemical Science Research Chair (ESRC), Chemistry Department, King Saud University, Riyadh, Kingdom of Saudi Arabia

*Address all correspondence to: mghanem@ksu.edu.sa

References

[1] Hua M, Zhang S, Pan B, Zhang W, Lv L, Zhang Q. Heavy metal removal from water/wastewater by nanosized metal oxides: A review. Journal of Hazardous Materials. 2012;**211-212**:317-331

[2] Fuchs V, Méndez L, Blanco M, Pizzio L. Mesoporous titania directly modified with tungstophosphoric acid: Synthesis, characterization and catalytic evaluation. Applied Catalysis A: General. 2009;**358**:73-78

[3] Wan Y, Yang H, Zhao D. Host-guest chemistry in the synthesis of ordered nonsiliceous Mesoporous materials. Accounts of Chemical Research. 2006;**39**:423

[4] Yang H, Zhao D. Synthesis of replica mesostructures by the nanocasting strategy. Journal of Materials Chemistry. 2005;**15**:1217-1231

[5] Blanco MN, Pizzio LR. Properties of mesoporous tungstosilicic acid/titania composites prepared by sol-gel method. Applied Surface Science. 2010;**256**:3546-3553

[6] Walcarius A. Mesoporous materials and electrochemistry. Chemical Society Reviews. 2013;**42**:4098-4140

[7] Choi J, Ide A, Truong YB, Kyratzis IL, Caruso RA. High surface area mesoporous titanium–zirconium oxide nanofibrous web: A heavy metal ion adsorbent. Journal of Materials Chemistry A. 2013;**1**:5847-5853

[8] Zhang R, Elzatahry AA, Al-Deyabb SS, Zhao D. Mesoporous titania: From synthesis to the application. Nano Today. 2012;**7**:344-366

[9] Collins PJ. Liquid Crystals: Nature's Delicate State of Matter. Princeton: Princeton University Press; 1990

[10] Ghanem MA. Electrochemical synthesis of nanostructured porous materials using liquid crystal and colloidal templates and their magnetic and optical properties [PhD thesis]. University of Southampton; 2002

[11] Attard GS, Glyde JC, Göltner CG. Liquid-crystalline phases as templates for the synthesis of mesoporous silica. Nature. 1995;**378**:366-368

[12] Attard GS, Bartlett PN, Coleman NBR, Elliott JM, Owen JR, Wang JHW. Mesoporous platinum films from lyotropic liquid crystalline phases. Science. 1997;**278**:838-840

[13] Bagshaw SA, Prouzet E, Pinnavaia TJ. Templating of mesoporous molecular sieves by nonionic polyethylene oxide surfactants. Science. 1995;**269**:1242-1244

[14] Kresge CT, Leonowicz ME, Roth WJ, Vartuli JC, Beck JS. Ordered mesoporous molecular sieves synthesized by a liquid-crystal template mechanism. Nature. 1992;**359**:710-712

[15] Martin CR. Nanomaterials: A membrane-based synthetic approach. Science. 1994;**266**:1961-1966

[16] Martin CR. Membrane-based synthesis of nanomaterials. Chemistry of Materials. 1996;**14**:1739-1746

[17] Huo Q, Margolese DI, Stucky GD. Surfactant control of phases in the synthesis of mesoporous silica-based materials. Chemistry of Materials. 1996;**8**(5):1147-1160

[18] Bartlett PN, Baumberg JJ, Birkin PR, Ghanem MA, Netti MC. Highly ordered macroporous gold and platinum films formed by electrochemical deposition through templates assembled from submicron diameter monodisperse polystyrene spheres. Chemistry of Materials. 2002;**14**(5):2199-2208

[19] Xia Y, Mokaya R. Hollow spheres of crystalline porous metal oxides: A generalized synthesis route via nanocasting with mesoporous carbon hollow shells. Journal of Materials Chemistry. 2005;**15**:3126-3131

[20] Wongsakulphasatch S, Kiatkittipong W, Saiswat J, Oonkhanond B, Striolo A, Assabumrungrat S. The adsorption aspect of Cu^{2+} and Zn^{2+} on MCM-41 and SDS-modified MCM-41. Inorganic Chemistry Communications. 2014;**46**:301-304

[21] Da'na E. Adsorption of heavy metals on functionalized-mesoporous silica: A review. Microporous and Mesoporous Materials. 2017;**247**(154):145-157

[22] Lee J-Y, Chen CH, Cheng S, Li HY. Adsorption of Pb(II) and Cu(II) metal ions on functionalized large-pore mesoporous silica. International journal of Environmental Science and Technology. 2016;**13**:65-76

[23] Pérez-Quintanilla D, Sánchez A, Sierra I. Preparation of hybrid organic-inorganic mesoporous silicas applied to mercury removal from aqueous media: Influence of the synthesis route on adsorption capacity and efficiency. Journal of Colloid and Interface Science. 2016;**472**:126-134

[24] Zhu W, Wang J, Wu D, Li X, Luo Y, Han C, et al. Investigating the heavy metal adsorption of mesoporous silica materials prepared by microwave synthesis. Nanoscale Research Letters. 2017;**12**:323

[25] Lin L, Thirumavalavan M, Lee JF. Facile synthesis of thiol-functionalized mesoporous silica-their role for heavy metal removal efficiency. Clean: Soil, Air, Water. 2015;**43**:775-785

[26] Wang C, Chen D, Jiao X. Lyotropic liquid crystal directed synthesis of nanostructured materials. Science and Technology of Advanced Materials. 2009;**10**:23001-23011

[27] Ernst S. Advances in Nanoporous Materials. Vol. 1. Amsterdam, Netherlands: Elsevier; 2009

[28] Yue Y, Gao Z. Synthesis of mesoporous TiO_2 with a crystalline framework. Chemical Communications. 2000;**18**:1755-1756

[29] Boettcher SW, Fan J, Tsung C-K, Shi Q, Stuky G. Harnessing the sol-gel process for the assembly of non-silicate mesostructured oxide materials. Accounts of Chemical Research. 2007;**40**:784-792

[30] Brinker CJ, Lu YF, Sellinger A, Fan HY. Evaporation-induced self-assembly: Nanostructures made easy. Advanced Materials. 1999;**11**(7):579-585

[31] Grosso D, Cagnol F, Soler-Illia GJDA, Crepaldi EL, Amenitsch H, Brunet-Bruneau A, et al. Fundamentals of mesostructuring through evaporation-induced self-assembly. Advanced Functional Materials. 2004;**14**(4):309-322

[32] Zhang Y, Zhao X-C, Wang Y, Zhou L, Zhang J, Wang J, et al. Mesoporous Ti–W oxide: Synthesis, characterization, and performance in selective hydrogenolysis of glycerol. Journal of Materials Chemistry A. 2013;**1**:3724-3732

[33] Miao Z, Song H, Zhao H, Xu L, Chou L. One-pot synthesis of mesoporous ZrPW solid acid catalyst for liquid phase benzylation of anisole. Catalysis Science & Technology. 2014;**4**:838-850

[34] Kuang IL, Hsueh Y-C, Chen H-S, Perng T-P. Mesoporous TiO_2/WO_3 hollow fibers with interior interconnected nanotubes for photocatalytic application. Journal of Materials Chemistry A. 2014;**2**:5387-5393

[35] Amphlet CB. Inorganic Ion Exchangers. Amsterdam: Elsevier; 1964

[36] Clearfield A, Nancollas GH, Blessing RH, Marinsky JA, Marcus Y. Ion Exchanger and Solvent Extraction. Vol. 5. New York: Marcel Dekker; 1973

[37] Clearfeld A. Inorganic ion exchangers with layered structures. Annual Review of Materials Science. 1984;**14**:205-229

[38] Mirajkar SP, Thangaraj A, Shiralkar VP. Sorption properties of titanium silicate molecular sieves. The Journal of Physical Chemistry. 1992;**96**:3073-3079

[39] Tegehall PE. Ion exchange on a titanium phosphate and formation of new crystalline phases by hydrolysis of the ion-exchanged phases. Acta Chemica Scandinavica. 1989;**43**:322-330

[40] Siddiqi ZM, Pathania D. Titanium(IV) tungstosilicate and titanium(IV) tungstophosphate: Two new inorganic ion exchangers. Journal of Chromatography. A. 2003;**987**:147-158

[41] Yavari R, Ahmadi SJ, Huang YD, Khanchi AR, Bagheri G, He JM. Synthesis, characterization and analytical application of a new inorganic cation exchanger-titanium (IV) molybdophosphate. Talanta. 2009;**77**:1179-1184

[42] Bennett WW, Teasdale PR, Panther JG, Welsh DT, Jolley DF. New diffusive gradients in a thin film technique for measuring inorganic arsenic and selenium(IV) using a titanium dioxide based adsorbent. Analytical Chemistry. 2010;**82**:7401-7407

[43] Yang P, Zhao D, Margolese DI, Chmelka BF, Stucky GD. Generalized syntheses of large-pore mesoporous metal oxides with semicrystalline frameworks. Nature. 1998;**396**:152-155

[44] Zhao D, Feng J, Huo Q, Melosh N, Fredrickson GH, Chmelka BF, et al. Triblock copolymer syntheses of mesoporous silica with periodic 50 to 300-angstrom pores. Science. 1998;**279**:548-552

[45] Yuan Q, Liu Y, Li LL, Li ZX, Fang CJ, Duan WT, et al. Highly ordered mesoporous titania-zirconia photocatalyst for applications in degradation of rhodamine-B and hydrogen evolution. Microporous and Mesoporous Materials. 2009;**124**:169

[46] Calleja G, Serrano DP, Sanz R, Pizarro P. Mesostructured SiO_2-doped TiO_2 with enhanced thermal stability prepared by a soft-templating sol-gel route. Microporous and Mesoporous Materials. 2008;**111**:429

[47] Sing KSW, Everett DH, Haul RAW, Moscou L, Pierotti RA, Rouquerol J, et al. Reporting physisorption data for gas/solid systems with special reference to the determination of surface area and porosity. Pure and Applied Chemistry. 1985;**57**:603-619

[48] Da'na E, Sayari A. Adsorption of heavy metals on amine-functionalized SBA-15 prepared by co-condensation: Applications to real water samples. Desalination. 2012;**285**:62-67

[49] Ezzeddine Z, Batonneau-Gener I, Pouilloux Y, Hamad H, Saad Z, Kazpard V. Divalent heavy metals adsorption onto different types of EDTA-modified mesoporous materials: Effectiveness and complexation rate. Microporous and Mesoporous Materials. 2015;**212**:125-136

[50] Hami DM, Yaftian MR, Pilehvari M, Rostamnia S. SBA-15 mesoporous materials decorated with organic ligands: Use as adsorbents for heavy metal ions. Journal of the Iranian Chemical Society. 2015;**12**:561-572. DOI: 10.1007/s13738-014-0513-8

[51] Anbia M, Kargosha K, Khoshbooei S. Heavy metal ions removal from aqueous media by modified magnetic mesoporous silica MCM-48. Chemical Engineering Research and Design. 2015;**93**:779-788

Chapter 2

Mesoporous Materials Prepared Using Cashew Nut Shell Liquid and Castor Oil as Surfactants

James Mgaya and Egid Mubofu

Abstract

Preparation of useful materials using renewable resources, which are not in competition with food production is of particular importance in the current efforts to replace non-renewable resources. One example of a potential renewable resource, which is attracting the attention of researchers in the preparation of useful materials is cashew nut shell liquid (CNSL). CNSL which is a by-product of cashew processing factories, is a mixture of four potential compounds, namely anacardic acid, cardanol, cardol and 2-methyl cardol. Among other potential applications, cashew nut shell liquid is a good template source for preparation of mesoporous materials. Heterogeneous catalysts prepared using CNSL templates are more efficient than those prepared using the commercially available templates. The pore sizes of mesoporous materials prepared using CNSL templates are large (up to 25 nm) enough to immobilize enzymes. Another renewable resource; castor oil, has also been reported to be a good template source for preparation of mesoporous materials. This chapter therefore is aimed at describing in detail the preparation, characterization and applications of mesoporous materials templated by cashew nut shell liquid and castor oil.

Keywords: cashew nut shell liquid, castor oil, micelle templated silica, heterogeneous catalysts, bio-catalysis

1. Introduction

1.1 Mesoporous materials

According to IUPAC, porous materials are divided into three classes, namely microporous (with a pore diameter below 2 nm), mesoporous (2–50 nm), and macroporous (larger than 50 nm) materials. The discovery in 1992 by Mobil researchers [1, 2] of the meso-structured MCM materials obtained by self-assembly of surfactants in aqueous solutions and silicates or aluminosilicates was one of the major breakthroughs in material science. Such materials have found great utility in sorption media, biosensors, biomedical and catalyst applications due to their large internal surface area. Their high density of silanol groups on the pore walls is beneficial to the introduction of functional groups with high coverage. One of the methods to synthesize mesoporous materials is to use a surfactant-templated sol-gel system [1, 2]. In the sol-gel technique, one starts with an appropriate alkoxide such as tetramethyl orthosilicate (TMOS) or tetraethyl orthosilicate (TEOS), which is mixed with water and a mutual solvent, such as ethanol or methanol, to form a solution, and the hydrolysis of the silicate leads to the formation of silanol groups, Si—OH. As the hydrolysis and condensation reactions (**Figure 1**) continue, viscosity increases until the sol ceases to flow and form the gel.

These silica materials prepared using sol-gel reactions with micelle templates are generally called micelle templated silica (MTS). The silica surface display different types of Si—O bonds, either silanols (≡Si—OH) or siloxanes (≡Si—O—Si≡), which have important influence on the particle properties.

1.2 Cashew nut shell liquid (CNSL) and castor oil as green templating agents

The general synthesis strategy of micelle templated silica (MTS) involves the condensation of an appropriate monomer in the presence of a templating agent (**Figure 2**). The templating agent is later removed from the composite material by soxhlet extraction, hot filtration or calcinations revealing a network of pores [3–5]. Surfactants are the commonly used template sources in the synthesis of micelle templated silica materials. They are amphiphilic compounds (usually organic compounds), which means they contain both hydrophobic groups (their tails) and hydrophilic groups (their heads). As a result of their amphiphilic nature, surfactants form aggregates (self-assemble) in solution mostly due to the hydrophobic effect. In the resultant aggregates, the amphiphilic molecules are arranged such that their hydrophobic groups are shielded from the aqueous medium by the polar head groups [6]. In dilute solution, surfactant molecules exist as monomers, but organize spontaneously into micelle when their concentration exceeds the critical micellar concentration [7]. Surfactant self-assembly (micelle) directs the organization of inorganic and organic precursors via non-covalent interactions: electrostatics, van der Waals, hydrogen bonding and π-π interactions.

The structure of the components of CNSL (cardanol, anacardic acid, cardol and 2-methyl cardol) contains polar phenolic part and non-polar 15 carbon alkyl non-polar chain which makes CNSL a suitable surfactant for preparation of mesoporous materials [8]. Likewise, the presence of carboxylic and hydroxyl groups in ricinoleic acid (which is a major component of castor oil; 90%) makes it an amphiphilic

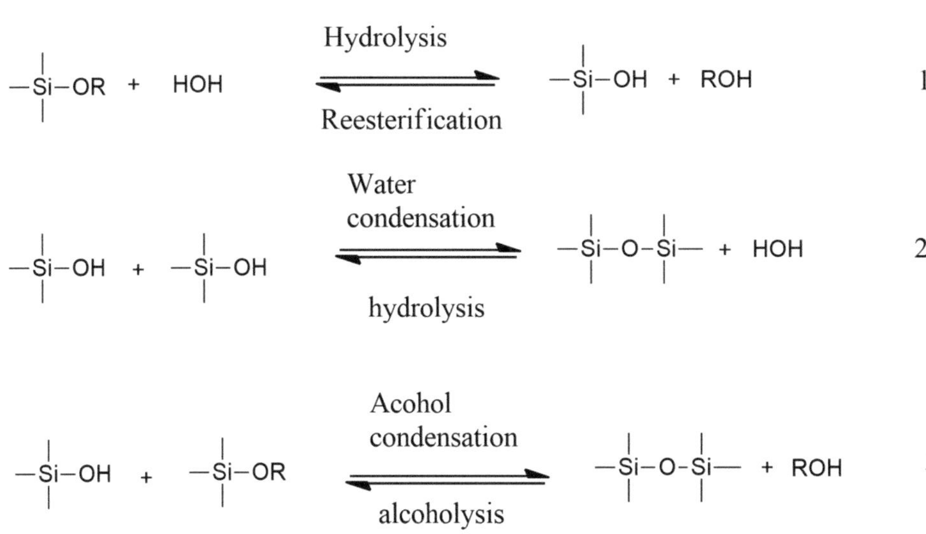

Figure 1.
Sol-gel reactions for the formation of silica gel.

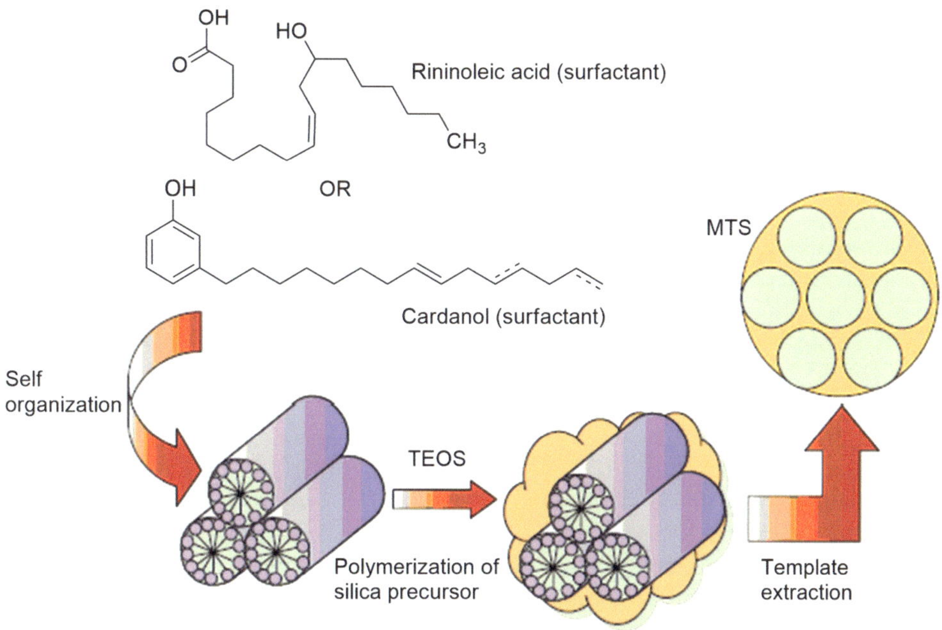

Figure 2.
Formation of micelle templated silica using CNSL and castor oil templates.

molecule in which the polar head is comprised of carboxylic acid and hydroxyl groups, while the rest of the carbon chain constitutes a hydrophobic tail. The amphiphilic nature of ricinoleic acid qualifies the castor oil as a source of micelles (template) for the synthesis of mesoporous materials.

The most commonly utilized groups of surfactants are the quaternary ammonium salts [9], the neutral amines [10] and the block copolymers [11]. It is proposed [12] that neutral surfactants direct formation of mesophase through hydrogen bonding interactions and self-assembly between neutral amine micelles and neutral inorganic precursors while charged surfactants involve electrostatic interactions. For the quaternary-derived materials, it is very difficult to remove the template in such a way that it can be recovered. In contrast, the removal of neutral surfactants (amines and polyesters) is very simple. The relatively weak interactions between the template and the wall of the material (hydrogen bonding) can be broken easily by heating in hydrogen bonding solvents such as ethanol and the template can be recovered completely, leaving behind the inorganic framework and a solution of the template [13], hence, this is the greenest way to produce mesoporous materials. Castor oil and CNSL are environmentally friendly and biocompatible neutral surfactant that have proven (as discussed in the next sections) to be potential replacement of commercially available surfactants which most of them are not environmentally friendly. Successful preparation of mesoporous silica materials using CNSL templates is one of great contributions of knowledge considering the fact that science has taken a new direction of research into using renewable raw materials in making industrial chemicals.

2. CNSL templated mesoporous silica materials (CNSL-MTS)

CNSL-MTS materials with functionalized surfaces for various applications can be prepared by either one pot co-condensation method or grafting method (also known as postsynthetic method). In co-condensation (direct or one-pot) synthesis, a condensable precursor bearing the desired functional group is added to the mixture containing the components for the formation of the mesoporous silica. In the postsynthetic method, the functional group is introduced after the formation of the mesoporous silica, either before or after removal of the template [14]. The one pot co-condensation method is the commonly used method of preparation of CNSL templated mesoporous materials. The synthesis is done by adding CNSL (surfactant) into a stirred aqueous ethanol (solvent) at ambient conditions of temperature and pressure. To this solution, TEOS (silica source) and a functionalizing agent (example; APTMS) are separately but simultaneously and rapidly added (in a ratio of 1:4 or 1:9). After 18 h, the resulting thick paste is vacuum filtered, thoroughly washed with ethanol. The template is removed by Soxhlet reflux using ethanol as a solvent for 10 h. The final solid (MTS) is dried in an oven at 100°C for 8 h ready for further derivatization and application.

Scanning electron microscopy (SEM) images have revealed that the CNSL-MTS particles are roughly spherical. The infrared spectra confirmed the successful functionalization of CNSL-MTS material with organic groups. For example, for amino functionalized samples, the peak at 1631 cm^{-1} which corresponds to the N—H bending implies successful functionalization with amino group. Nitrogen physisorption studies have further revealed that CNSL-MTS have large pore sizes of up to 25 nm, with surface area ranging from 100 to 214 m^2/g [15]. The recorded pore size of CNSL-MTS is by far larger than that of MTS materials synthesized using other neutral surfactants such as Igepal CA-720 (2.5–3.8 nm), Tergitol 15-S-12 (2.5–4.5), and Triton-X100 (2.0–60 nm) [16]. The pore size of CNSL-MTS is also larger than that of MTS prepared using nonionic block copolymer template (10 nm) [17], cationic surfactant (2.6–2.8 nm) [18], and also using a mixture of cationic and non-ionic surfactants (2.2–2.68 nm) [19]. However, the recorded surface area of CNSL-MTS (100–214 m^2/g) is smaller compared to those prepared using other neutral surfactants: Igepal CA-720 (up to 1400 m^2/g), Tergitol 15-S-12 (up to 1600 m^2/g), Triton-X100 (up to 1400 m^2/g). It is also less than that of MTS prepared using non-ionic block copolymer (770 m^2/g) [17], cationic surfactant (up to 1000 m^2/g) [18], and also using a mixture of cationic and non-ionic surfactants (1432 m^2/g) [19].

2.1 Applications of CNSL templated mesoporous silica materials

The pore size of the CNSL-MTS is large enough to immobilize large molecules such as invertase enzymes. They are also suitable for removal of heavy metals from aqueous solutions and for preparation of heterogeneous catalysts such as copper(II) Schiff base catalysts.

2.1.1 Immobilization of invertase enzymes on CNSL-MTS

Enzyme immobilization can be archived via several reported methods such as covalent attachment to the surface of water-insoluble material, entrapment inside a matrix or gel that is permeable to an enzyme, substrate and products, encapsulation and adsorption of an enzyme on a solid support [20, 21]. However, there is no single immobilization method which serves well for all enzymes. The activity of the immobilized enzymes depends on the immobilization technique used, conditions required for enzymatic treatment and the characteristics of the supporting materials. The type of immobilization method chosen determines the kind of modification that must be done on the surface of supporting material before immobilizing an enzyme.

Successful immobilization of invertase enzymes on CNSL-MTS via covalent attachment has been achieved [17]. CNSL-MTS material for covalent attachment of invertase enzyme is prepared by co-condensation of tetraethoxysilane (TEOS) and

3-aminopropyltrimethoxysilane (AMPTS) using cashew nut shell liquid (CNSL) template. The resulting materials are further modified by reacting them with glutaraldehyde resulting in Glu-MTS (**Figure 3**). Activation of CNSL-MTS with glutaraldehyde offers three different kinds of interactions with an enzyme: hydrophobic, anionic exchange and covalent [22].

Immobilization of invertase on Glu-MTS by covalent method is done by stirring invertase enzyme and supported glutaraldehyde (in a 1:1 ratio) in sodium acetate buffer (pH 4.7) for 5 h at room temperature followed by overnight incubation at 4°C (**Figure 4**).

The activity of free and CNSL-MTS immobilized invertase have been tested on hydrolysis reaction of sucrose (**Figure 5**) at different temperatures, pH and substrate concentrations [5].

Initially the activities both free and CNSL-MTS immobilized invertase enzymes increase with increasing temperature and both biocatalysts show a maximum activity at 40°C. The immobilized invertase maintains stability up to 45°C, whereas the free invertase declines sharply in activity with increase in temperature above 40°C. Thus, immobilization of invertase on CNSL-MTS slightly improves thermal stability of the biocatalyst. The optimum pH for free and CNSL-MTS immobilized invertase are 5.0 and 4.0 respectively. Under these optimum conditions of temperature and pH, the activities of free and immobilized enzymes are 15,229 U and 14,833 U respectively. This corresponds to reduction of approximately 3% (upon immobilization on CNSL-MTS) of the maximal activity of the free enzyme. This indicates that no significant loss in activity occurred during the process of attaching the enzyme by covalent binding to CNSL-MTS. Similar observation have been made in activities of invertase enzymes immobilized on mesoporous materials prepared using 1,3,5-trimethylbenze (TMB) and Pluronic P123 surfactants [23].

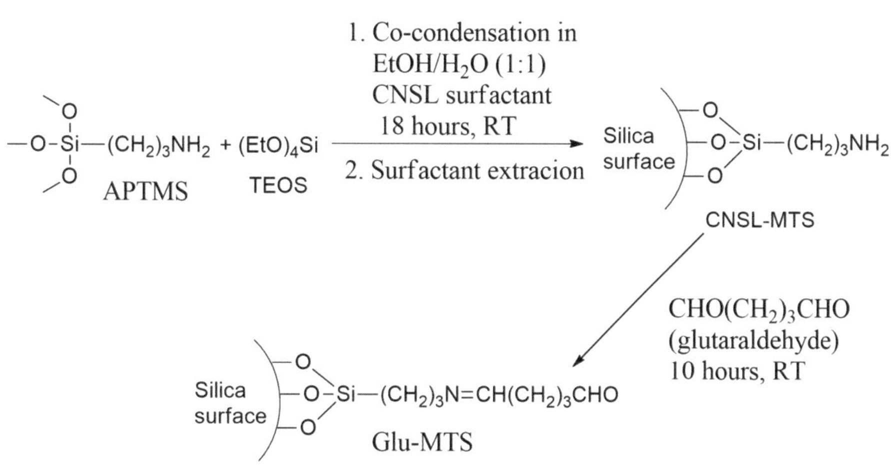

Figure 3.
Functionalization of CNSL-MTS with glutaraldehyde.

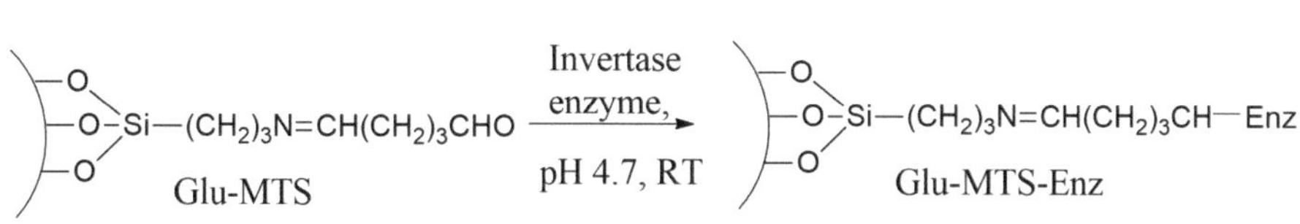

Figure 4.
Covalent binding of the invertase enzyme onto Glu-MTS.

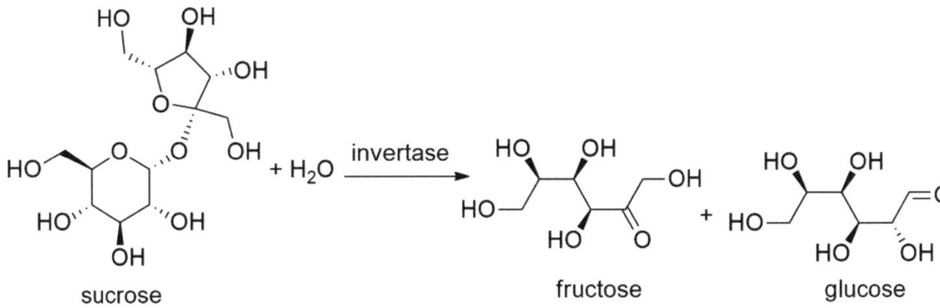

Figure 5.
Hydrolysis reaction of sucrose.

The activity of CNSL-MTS immobilized invertase is higher compared to the activity of invertase immobilized on MTS material prepared using a commercially available dodecylmine surfactant (DDA-MTS); (4750U, corresponding to reduction of approximately 70% of the maximal activity of the free invertase) under the same conditions of temperature and pH.

The reuse studies of the CNSL-MTS immobilized invertase enzyme done at 40°C and pH 4.7 at five-day intervals over a range of 30 days (used for up to 10 cycles in a day) revealed that the immobilized enzyme is highly stable and can remain active when stored for over 20 days. Therefore, although the activities of CNSL-MTS immobilized invertase enzyme seems slightly lower than of free invertase, immobilized invertase is capable of being reused for up to 10 cycles in a day without significant loss of activity making the overall activity of the immobilized invertase higher than that of free invertase. Comparing the reusability of CNSL-MTS (pore size is 25 nm) and DDA-MTS (pore size is 3 nm) immobilized invertase, it has been observed that DDA-MTS based invertase maintains its activities upon reuse whereas CNSL-MTS decreases in activity. This indicates that in DDA-MTS, the enzyme is attached on the outer surface whereas in CNSL-MTS the enzyme is immobilized inside the pores. This leads to pore blockage upon reuse which reduces the access of substrate to the active sites, consequently decreasing activity. The blockage of pores after reuse leads to a drop of surface area of a catalyst from 284 m^2/g before use to 84 m^2/g after use. Similar observations have been made elsewhere in non-enzymatic heterogeneous catalysis [24].

2.1.2 Immobilization of copper(II) Schiff base catalyst on CNSL-MTS

Preparation of CNSL-MTS material for immobilization of copper(II) Schiff base catalyst involves co-condensation of tetraethoxysilane and 3-aminopropyltriethoxysilane in the stirred mixture of CNSL (template source) and aqueous ethanol (solvent) at room temperature. The resulting aminopropyl-functionalized CNSL-MTS material (MTS-AMP) is then condensed with 2-hydroxybenzaldehyde to produce the corresponding Schiff bases. Finally, the MTS supported copper(II) catalyst is obtained by the reaction of the modified Schiff base support with a solution of copper(II) acetate in water, methanol or toluene (**Figure 6**) [25].

Characterizations by nitrogen physisorption and scanning electron microscopy (SEM) have revealed that the CNSL-MTS based heterogeneous catalysts are porous with pore diameters of up to 25 nm and grain sizes of up to 1.0 mm. The maximum copper loading of CNSL-MTS based heterogeneous catalysts is about 3% w/w which is higher when compared to heterogenous catalysts prepared using commercially available dodecylamine and hexadecylamine templates (2% w/w).

The activities and reusability of CNSL-MTS supported copper(II) Schiff base catalysts have been tested on the wet oxidation of maleic acid at room temperature using H_2O_2 as an oxidant [20]. Supported copper catalysts show high activity on various

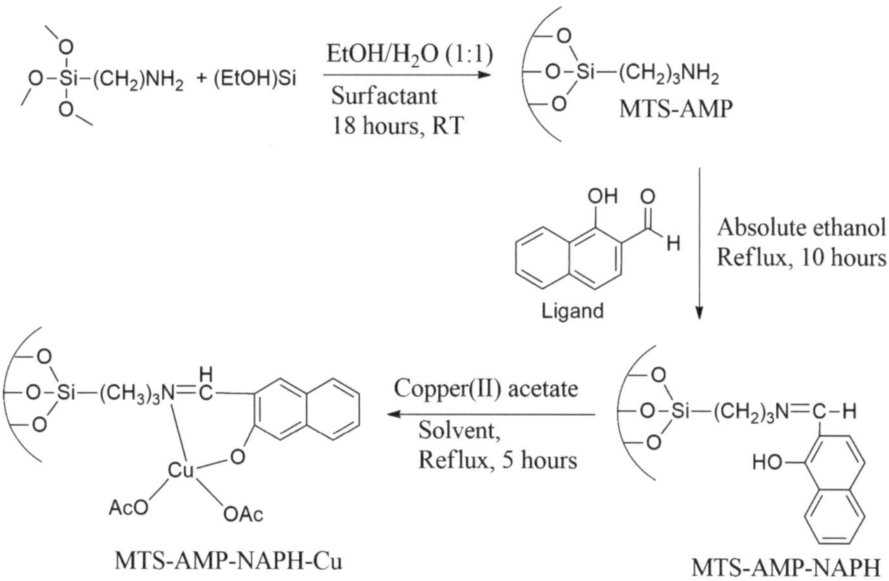

Figure 6.
Synthesis of MTS-AMP-NAPH-Cu catalyst.

oxidation reactions. For example, covalently supported copper-Schiff base complex of salicylaldehyde display an excellent catalytic activity in epoxidation reactions of various olefinic compounds using t-BuOOH as an oxidant [26]. Higher activity is also observed in wet oxidation of phenol using copper metal catalyst impregnated on MCM-41 [27].

Wet oxidation reaction of maleic acid produces carbon dioxide and water and its progress is monitored by using an HPLC technique. Although the reaction proceeds to some extent without a catalyst, it is highly accelerated by the copper catalyst. Catalysis by CNSL-MTS supported copper(II) Schiff base lead to conversion of about 95% yield after 40 min (with turn over number of about 1000) compared to less than 5% yield without catalyst after the same reaction time. Similar high activity is observed in decomposition of hydrogen peroxide (98% conversion after 50 min) using copper(II) Schiff base catalyst immobilized on SBA-15 supports (pore size = 8.8 nm, surface area = 346 m^2/g) [28]. Similar higher activity has also been reported in oxidation of sulfides to sulfoxides using H_2O_2 catalyzed by copper(II) Schiff base complex containing MCM-41 (up to 98% conversion) [29]. The interesting findings on these reported findings are not only the higher activity of the heterogeneous catalysts, but also their ability to be reused for up to five cycles without significant loss of activity.

2.1.3 Removal of lead(II) ions from aqueous solutions using CNSL-MTS

CNSL-MTS materials are modified with thiol functionality to suit lead metal ion adsorption. One way to synthesize thiol modified CNSL-MTS material is via a co-condensation of TEOS and (3-mercaptopropyl)-trimethoxysilane (MPTMS) using CNSL template. A typical synthesis involves adding TEOS and MPTMS (in 1:4, or 1:9) separately but simultaneously to a stirred solution of aqueous of ethanol and CNSL. The reaction is allowed to proceed for 18 h to obtain a thick solution which is filtered and air dried for 24 h. Finally, the template is removed by Soxhlet extraction using ethanol solvent.

Lead(II) adsorption isotherms are obtained by equilibrating CNSL-MTS samples with aqueous solution of $Pb(NO_3)_2$ at different metal concentrations, while maintaining a pH = 4 which is below the precipitation level of lead (pH > 6.5). After equilibration, the sample is filtered and the Pb(II) ion concentration is measured using atomic absorption spectrophotometer. The maximum adsorption capacity observed for CNSL-MTS is 66.7 mg Pb(II) per g adsorbent. Higher adsorption of Pb(II) ions (272 mg Pb(II) per g adsorbent) using mesoporous materials with P-Containing pendant groups is reported elsewhere [30].

2.2 Synthesis of organoamine-silica hybrids using CNSL components as templates

Cashew nut shell liquid has four major components as described in the introduction section of this chapter. Investigation on which of the major component of CNSL is responsible for formation of CNSL-MTS has been done [31]. The findings show that, of the major components of CNSL cardanol is the efficient template source yielding 66% mesoporous organoamine-silica compared to anacardic acid (6%). Cardanol templated organoamine silica materials show high activity when tested with Henry reaction (as a model base catalyzed reaction) giving 81–98% yield.

3. Castor oil templated mesoporous silica materials (CO-MTS)

The preparation procedure of castor oil templated silica (CO-MTS) is more or less similar to that used in preparation of CNSL-MTS discussed in Section 2. It involves condensing a silica source such as TEOS in a stirred mixture of castor oil (surfactant) and aqueous ethanol (solvent) in acidic (HCl) medium. The reaction is allowed to proceed for 18 h. The resulting product mixture is filtered to remain with the solid which is subjected to Soxhlet extraction using ethanol as extracting solvent to give mesoporous silica (CO-MTS). Nitrogen physisorption studies have revealed that castor oil templated MTS contains: 6.7 nm pore diameter, 899 m^2/g surface area and 1.1 cm^3/g total pore volume [32]. They exhibit type IV adsorption-desorption isotherms which is characteristic of mesoporous structures. SEM analysis have reveled clusters of spherical shaped materials which suggest that growth of silica occurs within colloidal dispersions [33].

3.1 Applications of castor oil templated silica

The pore size of CO-MTS is small when compared to that of CNSL-MTS (pore diameter = 25 nm), however, they are large enough for such applications as catalysis, metal adsorptions, and immobilization of biomolecules [34, 35].

3.1.1 Amine functionalization of CO-MTS for the transesterification of tributyrin with methanol

One of the unique features of mesoporous silica material is that they provide a room for surface modification to suit a particular application. CO-MTS materials modified with Amino groups are potential base catalysts used in esterification reactions [36, 37]. Preparation of amine modified CO-MTS has been achieved via grafting method [33]. It involves dispersing the oven dried samples of CO-MTS in dried toluene followed by addition of amines such as RNH_2, NN, or DN. The resulting amine functionalized CO-MTS are termed in this case as NH_2-MTS, NN-MTS and DN-MTS respectively (**Figure** 7). Surface analysis by XPS of the chemical composition on the silica grafted groups has revealed successful attachment of N containing groups for all samples. The surface area of the CO-MTS decreases from 899 m^2/g to 510,407 and 246 m^2/g upon modification with RNH_2, NN, and DN, respectively. This indicates more significant blockage as the bulkiness of the organoamines increases [29].

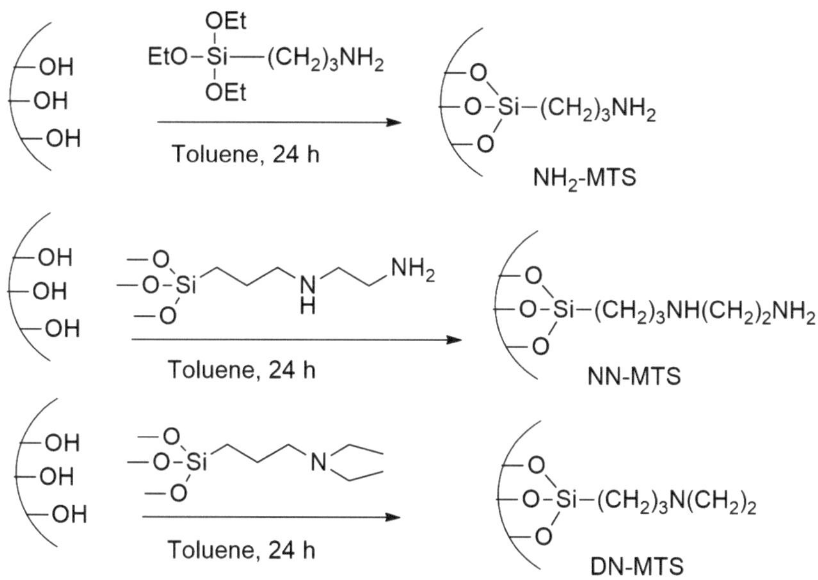

Figure 7.
Grafting amines to the MTS.

The activity of amine functionalized CO-MTS investigated on transesterification of tributyrin at 60°C showed to follow the following conversion trend: NH_2-MTS (8.9%) < NN-MTS (22%) < DN-MTS (26.7%). The trend suggests that there is an impact of amine functionality on transesterification activity and shows the superior performance of the tertiary amine. The catalysts have turnover frequencies of 0.9, 6.75 and 1.45 h^{-1} for NH_2-MTS, NN-MTS and DN-MTS catalysts respectively. The yields are relatively low compared to some reported esterification reaction (yields of above 40%) using other amine-functionalized ordered mesoporous silica, however different esterification substrates were used [38, 39].

3.1.2 Immobilization of trypsin enzymes on CO-MTS via thiol-disulfide interchange

Immobilization of trypsin on CO-MTS can be achieved via adsorption (modification is not needed) or via covalent attachment (modification is required). Immobilization of trypsin on CO-MTS has been achieved via thiol-disulfide interchange (covalent attachment). Thiol modified CO-MTS is prepared by co-condensation of 3-mercaptopropyltriethoxysilane and tetraethylorthosilicate in a 1:4 molar ratio respectively using castor oil template. Nitrogen physisorption studies have revealed that, the pore volume of thiol modified CO-MTS materials decreases from 1.0336 cm^3/g before immobilization to 0.8169 cm^3/g after trypsin immobilization [32]. The decrease in pore volume after enzyme immobilization is not strange and is due to coating of the support surface by the enzyme molecule [40].

The activity of both free and immobilized trypsin is tested on hydrolysis of N-α-benzoyl-D,L-arginine-p-nitroanilide (BAPNA). Hydrolysis of BAPNA by trypsin (**Figure 8**) cleaves the bond between the arginine and the p-nitroaniline to produce free p-nitroaniline, which is yellow and easily measured in a UV-VIS spectrophotometer.

Both free and immobilized trypsin show maximum activity on BAPNA hydrolysis at a temperature range of 35–38°C. CO-MTS immobilized trypsin show some stability up to 45°C while free trypsin has a sharp decrease on activity at that temperature. The V_{max} for free and CO-MTS immobilized trypsin are 0.32, and 0.182 U, respectively.

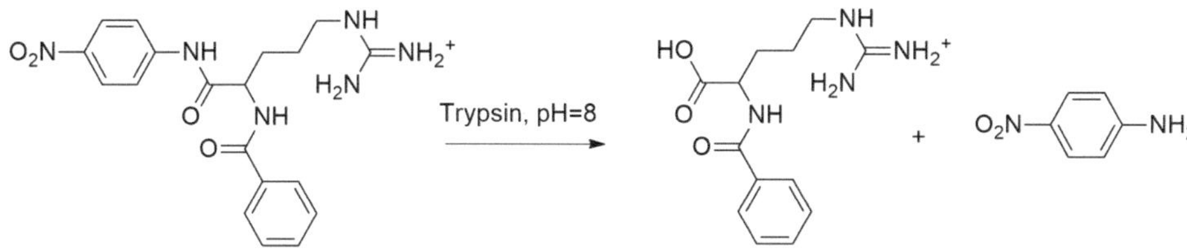

Figure 8.
Hydrolysis of BAPNA substrate for enzymatic activity assay of trypsin.

The values are not far from those reported elsewhere [35]. The immobilized trypsin is capable of being reused up to four cycles with slight decrease on its activity [32].

3.1.3 Immobilization of palladium catalysts onto CO-MTS

The CO-MTS material for immobilization of palladium catalyst is firstly modified with amino groups via the co-condensation reaction of tetraethyl orthosilicate (TEOS) and 3-aminopropyltriethoxysilane (AMPS). The reaction is carried out in ethanol/water solvent using castor oil surfactants at room temperature. The prepared aminopropyl functionalized CO-MTS is further modified by reacting with quinoline-2-carboxaldehyde to obtain CO-MTS supported Schiff base. Immobilization of palladium catalyst is achieved by reacting the modified Schiff base support (MTS-AMP-Q), with palladium (II) chloride solution in toluene (**Figure 9**) [41].

The CO-MTS supported catalysts have been tested for activity in the wet oxidation of a mixture of phenol, ocresol and resorcinol using hydrogen peroxide oxidant. The conversion rate of the tested phenols follows the trend: resorcinol > phenol > o-cresol. The catalysts are active from room temperature to 70°C, but are slightly more effective when the reaction temperature is greater than room temperature. For instance, in 60 min of reaction, the catalyst achieves about 50% of phenolic degradation at room temperature and about 55% conversion at 70°C. Other reports have shown even better phenol conversion of up to 100% [42–44].

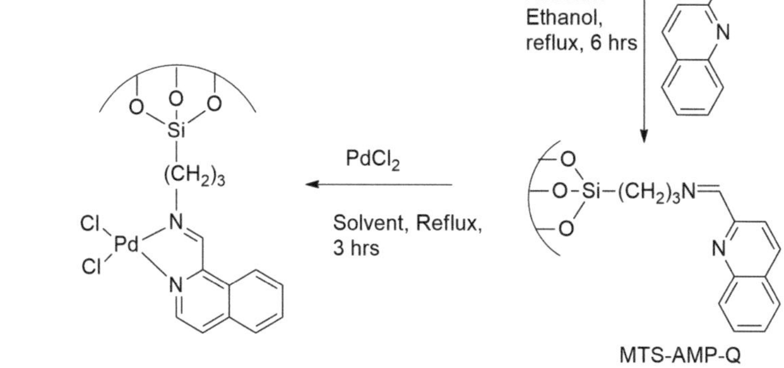

Figure 9.
Stepwise preparation of CO-MTS supported palladium catalyst.

4. Sunflower oil templated mesoporous silica (SO-MTS)

Sunflower oil has also been reported to be a potential renewable surfactant for synthesis of mesoporous materials. However, sunflower is an edible oil and therefore its use as raw material in industries is highly likely to face competition from food production. Mesoporous materials prepared using sunflower oil have pore diameter of up to 10.1 nm with surface area of 734.1 m^2/g [45].

4.1 Adsorption studies of Cr(III) using cyano functionalized SO-MTS

Preparation of cyano functionalized SO-MTS materials is achieved by one pot co-condensation of 2-cyanoethyltriethoxysilane (CETS) and tetraethoxysilane (TEOS) using sunflower oil as templating agent. The adsorption sites are either the tethered cyano groups or carboxylic groups obtained by hydrolysis of cyano groups (**Figure 10**).

The functionalized SO-MTS materials have adsorption maximum of 19.7 mg Cr(III) ions per g of adsorbent. Adsorption studies of Cr(III) ions from tannery wastewater have revealed that functionalized SO-MTS material can remove Cr(III) ions up to 83% [27].

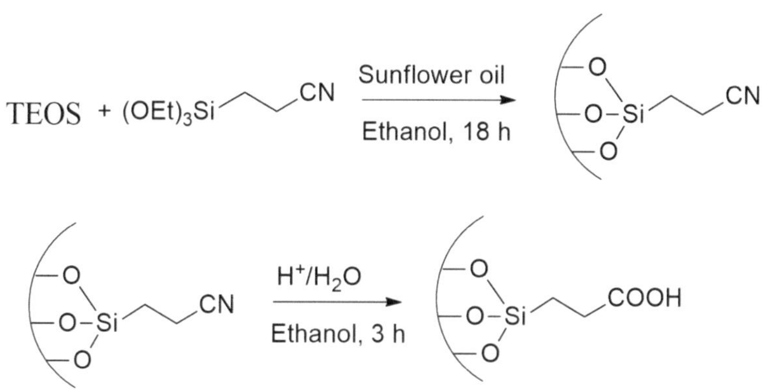

Figure 10.
Synthesis of sunflower oil templated MTS-CN and MTS-COOH.

5. Conclusion

Details of routes to synthesis of mesoporous silica materials using renewable surfactants as potential sources of templates are well covered. Cashew nut shell liquid, castor oil and sunflower oil have all shown to be good bio-based sources of templates for synthesis of mesoporous silica materials. The prepared materials using these bio-based surfactants have surface properties desirable for potential applications such as enzyme immobilization, preparation of heterogenous copper(II) Schiff base catalysts and removal of heavy metals by adsorption. Heterogenous catalysts prepared using these bio-based surfactants have shown high activities, reusability, and stability. Heavy metal removal by adsorption of up to 80% has been archived.

Conflict of interest

There are no conflicts of interest.

Author details

James Mgaya[1*] and Egid Mubofu[2]

1 University of Dar es Salaam (DUCE), Dar es Salaam, Tanzania

2 University of Dodoma, Dodoma, Tanzania

*Address all correspondence to: jamesmgaya@yahoo.com

References

[1] Beck JS, Vartuli JC, Roth WJ, Leonowicz ME, Kresge CT, Schmitt KD, et al. A new family of mesoporous molecular sieves prepared with liquid crystal templates. Journal of the American Chemical Society. 1992;**114**(27):10834-10843. DOI: doi/abs/10.1021/ja00053a020

[2] Kresge CT, Leonowicz ME, Roth WJ, Vartuli JC, Beck JC. Ordered mesoporous molecular sieves synthesized by a liquid-crystal template mechanism. Nature. 1992;**359**:710-712

[3] Wu SH, Lin HP. Synthesis of mesoporous silica nanoparticles. Chemical Society Reviews. 2013;**42**(9):3862-3875

[4] Naik B, Ghosh N. A review on chemical methodologies for preparation of mesoporous silica and alumina based materials. Recent Patents on Nanotechnology. 2009;**3**(3):213-224

[5] Mehmood A, Ghafar H, Yaqoob S, Gohar UF, Ahmad B. Mesoporous silica nanoparticles: A review. Journal of Developing Drugs. 2017;**06**(2):1-14

[6] Borisov OV, Zhulina EB, Leermakers FAM, Müller AHE. Transcatheter arterial embolization for treating intractable post-partum hemorrhage: Early success indicators. Chinese Journal of Radiology. 2009;**34**(4):253-261

[7] Zdziennicka A, Szymczyk K, Krawczyk J, Jańczuk B. Critical micelle concentration of some surfactants and thermodynamic parameters of their micellization. Fluid Phase Equilibria. 2012;**322-323**:126-134

[8] Nagarajan R. Molecular packing parameter and surfactant self-assembly: The neglected role of the surfactant tail. Langmuir. 2002;**18**:31-38

[9] Wang P, Chen S, Zhao Z. Synthesis of ordered porous SiO_2 with pores on the border between the micropore and mesopore regions using rosin-based quaternary ammonium. RSC Advances. 2015;**5**:11223-11228. DOI: doi.org/10.1039/C4RA12113A

[10] Prado AGS, Airoldi C. Different neutral surfactant template extraction routes for synthetic hexagonal mesoporous silicas. Journal of Materials Chemistry. 2002;**12**:3823-3826. DOI: 10.1039/b204060c

[11] Feng P, Bu X, Pine DJ. Control of pore sizes in mesoporous silica templated by liquid crystals in block copolymer-cosurfactant-water systems. Langmuir. 2000;**23**:5304-5310

[12] Tanev PT, Pinnavaia TJ. A neutral templating route to mesoporous molecular sieves. Science. 1995;**267**(5199):865-867

[13] Clark J, Macquarrie D, editors. Handbook of Green Chemistry and Technology. Blackwell Science Ltd; Oxford, UK 2002. 120-127 p

[14] Brühwiler D. Postsynthetic functionalization of mesoporous silica. Nanoscale. 2010;**2**(6):887-892

[15] Mubofu EB, Mdoe JEG, Kinunda G. The activity of invertase immobilized on cashew nut shell liquid-templated large pore silica hybrids. Catalysis Science & Technology. 2011;**1**:1423-1431. DOI: 10.1039/c1cy00033k

[16] Richer R, Mercier L. Direct synthesis of functional mesoporous silica by neutral pH nonionic surfactant assembly: Factors affecting framework structure and composition. Chemistry of Materials. 2001;**13**(9):2999-3008

[17] Chen D, Li Z, Yu C, Shi Y, Zhang Z, Tu B, et al. Nonionic block copolymer and anionic mixed surfactants directed synthesis of highly ordered

mesoporous silica with bicontinuous cubic structure. Chemistry of Materials. 2005;**17**(12):3228-3234

[18] Che S, Li H, Lim S, Sakamoto Y, Terasaki O, Tatsumi T. Synthesis mechanism of cationic surfactant templating mesoporous silica under an acidic synthesis process. Chemistry of Materials. 2005;**17**(16):4103-4113

[19] Song MG, Kim JY, Cho SH, Kim JD. Mixed cationic-nonionic surfactant templating approach for the synthesis of mesoporous silica. Langmuir. 2002;**18**(16):6110-6115

[20] Yamada K, Nakasone T, Nagano R, Hirata M. Retention and reusability of trypsin activity by covalent immobilization onto grafted polyethylene plates. Journal of Applied Polymer Science. 2003;**89**:3574-3581

[21] Costa SA, Azevedo HS, Reis RL. Enzyme immobilization in biodegradable polymers for biomedical applications. Journal of Materials Science. Materials in Medicine. 2005;**15**(4):301-324. DOI: 10.1201/9780203491232.ch17

[22] Ribeiro LMO, Falleiros LNSS, de Resende MM, Ribeiro EJ, Almeida RMRG, da Silva AOS. Immobilization of the enzyme invertase in SBA-15 with surfaces functionalized by different organic compounds. Journal of Porous Materials. 2018;**0**(0):1-13. DOI: doi.org/10.1007/s10934-018-0615-2

[23] Szymańska K, Bryjak J, Jarzębski AB. Immobilization of invertase on mesoporous silicas to obtain hyper active biocatalysts. Topics in Catalysis. 2009;**52**(8):1030-1036

[24] Clark JH, Macquarrie DJ, Mubofu EB. Preparation of a novel silica-supported palladium catalyst and its use in the Heck reaction. Green Chemistry. 2000;**2**(2):53-56. DOI: 10.1039/A908685D

[25] Hamad FB, Mubofu EB, Makame YMM. Wet oxidation of maleic acid by copper(II) Schiff base catalysts prepared using cashew nut shell liquid templates. Catalysis Science & Technology. 2011;**1**(3):444-452. DOI: 10.1039/c1cy00001b

[26] Jana S, Dutta B, Bera R, Koner S, Jada V, Uni V. Anchoring of copper complex in MCM-41 matrix : A highly efficient catalyst for epoxidation of olefins by tert-BuOOH. Langmuir. 2007;**12**:2492-2496

[27] Wu Q , Hu X, Lock P, Song X, Qing G. Copper/MCM-41 as catalyst for the wet oxidation of phenol. Applied Catalysis B: Environmental. 2001;**32**:151-156

[28] Rajabi F, Pinilla-de Dios M, Luque R. Highly ordered nanomaterial functionalized copper Schiff base framework: Synthesis, characterization, and hydrogen peroxide decomposition performance. Catalysts. 2017;7:216

[29] Mahdavi H, Nikoorazm M, Ghorbani-Choghamarani A, Arshadi S. Synthesis and characterization of copper(II) Schiff base complex containing MCM-41 as robust and clean nano reactor catalyst for the selective oxidation of sulfides to sulfoxides using H_2O_2. Journal of Porous Materials. 2016;**23**(1):75-82

[30] Gunathilake C, Kadanapitiye MS, Dudarko O, Huang SD, Jaroniec M. Adsorption of lead ions from aqueous phase on mesoporous silica with P-containing pendant groups. ACS Applied Materials & Interfaces. 2015;**7**(41):23144-23152

[31] Msigala SC, Mdoe JEG. Synthesis of organoamine-silica hybrids using cashew nut shell liquid components as templates for the catalysis of a model henry reaction. Tanzania Journal of Science. 2012;**38**(1):24-34

[32] Mgaya J. Preparation of castor oil porous organosilica composites for immobilization of trypsin enzyme preparation of castor oil porous organosilica [thesis]. University of Dar es salaam; 2012

[33] Elimbinzi E, Nyandoro SS, Mubofu EB, Osatiashtiani A, Jinesh C. Synthesis of amine functionalized mesoporous silicas templated by castor oil for transesterification. MRS Advances. 2018;**3**(38):2261-2269. DOI: 10.1557/adv.2018.347

[34] Jang S, Kim D, Choi J, Row K, Ahn W. Trypsin immobilization on mesoporous silica with or without thiol functionalization. Journal of Porous Materials. 2006;**13**:385-391. DOI: 10.1007/s10934-006-8035-0

[35] Karimi B, Emadi S, Safari A, Kermanian M. Immobilization, stability and enzymatic activity of albumin and trypsin adsorbed onto nanostructured mesoporous SBA-15 with compatible pore sizes. RSC Advances. 2014;**4**:4387-4394. DOI: 10.1039/c3ra46002a

[36] Narasimharao K, Lee A, Wilson K. Catalyts i producton of biodiesel: A review. Journal of Biobased Materials and Bioenergy. 2007;**1**(1):1-12. DOI: 10.1166/jbmb.2007.002

[37] Zhang L, Guo W, Liu D, Yao J, Ji L, Xu N, et al. Low boiling point organic amine-catalyzed transesterification for biodiesel production. Energy and Fuels. 2008;**22**(2):1353-1357

[38] Guerrero VV, Shantz DF. Amine-functionalized ordered mesoporous silica transesterification catalysts. Industrial and Engineering Chemistry Research. 2009;**48**(23):10375-10380

[39] Liu Y, Zhao G, Zhu W, Wang J, Liu G, Zhang W, et al. Comparative study on the catalytic properties of amino-functionalized silica materials for the transesterification of dimethyl oxalate with phenol. Journal of the Brazilian Chemical Society. 2010;**21**(12):2254-2261

[40] Matte CR, Bordinhão C, Poppe JK, Benvenutti EV, Costa TMH, Rodrigues RC, et al. Physical-chemical properties of the support immobead 150 before and after the immobilization process of lipase. Journal of the Brazilian Chemical Society. 2017;**28**(8):1430-1439

[41] Andrew M, Mubofu EB. Catalytic efficiency of palladium(II) supported on castor oil based micelle templated silica for peroxide promoted catalytic wet oxidation of phenols. IRJPAC. 2017;**15**(1):1-12. DOI: 10.9734/IRJPAC/2017/36966

[42] Pestunova OP, Ogorodnikova OL, Parmon VN. Studies on the phenol wet peroxide oxidation in the presence of solid catalysts. Chemistry for Sustainable Development. 2003;**11**:227-232

[43] Britto JM, de Oliveira SB, Rabelo D, Rangel MC. Catalytic wet peroxide oxidation of phenol from industrial wastewater on activated carbon. Catalysis Today. 2008;**133-135**(1-4):582-587

[44] Feng YB, Hong L, Liu AL, Chen WD, Li GW, Chen W, et al. High-efficiency catalytic degradation of phenol based on the peroxidase-like activity of cupric oxide nanoparticles. International journal of Environmental Science and Technology. 2013;**12**(2):653-660

[45] Gervas C, Mubofu EB. Functionalized mesoporous organo-silica nanosorbents for removal of chromium (III) ions from tanneries wastewater. Journal of Porous Materials. 2015;**23**(1):83-93. DOI: 10.1007/s10934-015-0058-y

Section 2

Mesoporous Materials and their Potential Applications

Chapter 3

Synthesis of MCM-41/ZIF-67 Composite for Enhanced Adsorptive Removal of Methyl Orange in Aqueous Solution

Ratna Ediati, Pramita Elfianuar, Eko Santoso, Dety Oktavia Sulistiono and Muhammad Nadjib

Abstract

ZIF-67 and MCM-41/ZIF-67 composites were successfully synthesized with water solvent at room temperature. The amounts of MCM-41 added during synthesis were varied at 2.5, 5 and 10 (%w/w) toward the amount of ZIF-67, and the obtained solids were denoted as MC (2.5)/ZIF-67, MC (5)/ZIF-67, and MC (10)/ZIF-67, respectively. The X-ray diffraction (XRD) patterns of ZIF-67 and the composites showed characteristic peaks at 2θ of 7.32, 10.36, 12.69, 14.66, and 16.40°, similar to that of reported ZIF-67. The Fourier transform infra-red (FT-IR) spectra of all solids showed absorption bands at the same wavenumbers as reported for ZIF-67. The results of surface morphology analysis using scanning electron microscope (SEM) have shown that ZIF-67 and the composites have a cube shape, which is characteristic for the ZIF-67 standard. N_2 adsorption-desorption data showed that the specific surface area of ZIF-67 and MC (5)/ZIF-67 were 1079.2 and 1011.2 m^2/g, respectively, lower than that of MC (10)/ZIF-67 (1250.6 m^2/g). However, results of thermal gravimetric analysis (TGA) showed that the thermal stability of MC (10)/ZIF-67 reached 357°C, higher than that of ZIF-67 (325°C). Performance of the composites as adsorbent of methyl orange (MO) in aqueous solution showed that the MC (5)/ZIF-67 had the highest adsorption capacity of 167.635 mg/g, and followed the pseudosecond-order adsorption kinetics and Langmuir isothermal adsorption.

Keywords: ZIF-67, composite, MCM-41/ZIF-67, adsorption, methyl orange

1. Introduction

Organic dyes are pollutants that can provide toxic effects on microbial populations and can be toxic or carcinogenic to organisms and mammals [1]. Azo dyes are a group of synthetic organic dyes, which are most widely used in the textile, cosmetics, and food coloring industries, so they are found in industrial waste water. The release of waste water containing dyes, with different chemical compositions and toxicity, into waterways can cause environmental damage [2]. Methyl orange (MO) is one of the cationic dyes, which are widely used in the coloring industry. If the MO concentration in wastewater reaches 500 ppm, it can cause serious health problems, such as tachycardia, vomiting, cyanosis, jaundice, quadriplegia, and tissue necrosis. Therefore, removing MO in wastewater is needed to avoid the negative effects that arise [3–5].

Several methods have been reported to remove organic dyes in water, such as adsorption with solid materials [6], chemical and biological degradation [7, 8], electrochemical degradation [9], and ozonation [10]. Adsorption, however, is considered as a simple and effective method. The adsorbents used, such as activated carbon, agricultural waste, and chitosan [11], have shown low adsorption capacity and selectivity [12].

Metal organic frameworks (MOF) are a new class of nanoporous crystalline materials consisting of metal ions or clusters that are interconnected with organic ligands and described as materials with microstructure (pore size < 2 nm), adjustable pore size, high surface area (up to 5900 m^2/g), and high pore volume (up to 2 cm^3/g) [13–15]. With the advantages, MOF has been considered as a potential material for various applications, including gas storage, separation, adsorption, drug delivery, and catalysts [15]. As an adsorbent, MOF has been widely used to remove toxic metals and organic pollutants from aqueous solutions [16].

Zeolitic imidazolate frameworks (ZIF) are parts of MOF, composed of tetrahedral metal ions such as Zn^{2+} or Co^{2+}, which are bridged by imidazolate ligands [17, 18]. ZIF-67 ($Co(Hmim)_2$), with a sodalite (SOD) topology [18], has been evaluated for its performance as an adsorbent in an aqueous environment and shows a high adsorption capacity against several pollutants including azo dyes [12, 19]. In the adsorption process, the pore size of adsorbent is very influential. MOF with pore size of less than 2 nm (microporous) may cause diffusion problems toward large molecules. Mesoporous silica such as MCM-41, MCM-48, and SBA-15 have been reported as supports in MOF synthesis, as an effort to increase pore size of MOF and to reduce diffusion problems when used as adsorbents. Furthermore, mesoporous silica also has high thermal stability, surface area, and porosity [20, 21].

Based on the description above, in this study, removal of MO in aqueous solution was carried out using ZIF-67 and MCM-41/ZIF-67 composites. ZIF-67 and the composites were synthesized following the method reported by Gross et al. with a slight modification [22]. The experiments were carried out at room temperature with $CoCl_2{\cdot}6H_2O$ as a metal source and 2-methylimidazole as a ligand in aqueous solution. A certain amount of MCM-41 was added directly into the reaction mixture.

2. Experimental

2.1 Materials

Materials used in this study were cobalt chlorida hexahydrate ($CoCl_2.6H_2O$, Sigma Aldrich, 99%), 2-methylimidazole ($C_4H_6N_2$, Sigma Aldrich, 99%), deionized water, triethylamine (TEA), methanol (MeOH, Merck 99.8%), and methyl orange (Sigma Aldrich, 85%). Mesoporous silica (MCM-41) was synthesized using a procedure reported by Badamali et al. [23].

2.2 Synthesis of ZIF-67 and composite MCM-41/ZIF-67

ZIF-67 and MCM-41/ZIF-67 composites were synthesized with a molar ratio of metal:ligand of 1:11.6. The synthesis of ZIF-67 was begun by dissolving 6.09 g of 2-methylimidazole (MeIM) in 13.5 mL deionized water. Triethylamine (TEA) added was 0.5 mL and the amount of MCM-41 added was varied at 2.5, 5 and 10 w/w% toward the amount of the obtained ZIF-67. Then, $CoCl_2{\cdot}6H_2O$ solution was made by dissolving 1.52 g of $CoCl_2{\cdot}6H_2O$ in 10 mL deionized water. $CoCl_2{\cdot}6H_2O$ solution was mixed with 2-methylimidazole solution in a vial bottle and stirred for 20 min. Furthermore, the reaction mixture was centrifuged at 3000 rpm for 30 min. The obtained solids were washed with deionized water, followed by washing with methanol. Finally, the solids were dried in an oven at a temperature of 120°C for 16 hours.

2.3 Characterization ZIF-67 and MCM-41/ZIF-67 composites

The crystallinity of ZIF-67 and MCM-41/ZIF-67 composites were determined by XRD with CuKα radiation (λ = 1.5406 Å), with 40 kV accelerated voltage, and a current of 30 mA. The 2θ angle was recorded from 3 until 50°. Analysis of the existence of functional groups on ZIF-67 and MCM-41/ZIF-67 composites was carried out by FTIR spectrophotometer at wavenumbers of 4000–400 cm^{-1}. Analysis by scanning electron microscope (SEM) has been done to determine the surface morphology of synthesized material. TEM images were obtained using a JEOL JEM1400 transmission electron microscope at an accelerating voltage of 200 kV. Measurement of surface area and pore size distribution of the materials were carried out using surface area analyzer based on adsorption-desorption isotherm using nitrogen gas, and the samples were degassed at 150°C for 5 hours (3×10^{-3} Torr) before being analyzed. Thermogravimetric analysis (TGA) has been used to determine the thermal stability, and the samples were heated from temperature of 35–700°C with an air gas flow.

2.4 Adsorption experiment

Methyl orange (MO) adsorption test on the effect of contact time was carried out using 10 mg of adsorbent in 100 mg/L of MO solution. The mixture was stirred at 300 rpm for 5 until 40 min. In addition, the adsorption test was also carried out on the effect of the initial concentration and concentration of MO was varied from 50 to 330 mg/L. After adsorption process, the solution was centrifuged at a rate of 3500 rpm for 10 min to separate the adsorbent from the solution. Concentration of MO was measured using a UV-Vis spectrophotometer at the maximum wavelength. Adsorption capacities of adsorbent and %removal of MO were calculated using the formula:

$$Q_e = \frac{(C_o - C_e) \times V}{W} \tag{1}$$

$$\%Removal = \frac{C_0 - C_e}{C_o} \times 100 \tag{2}$$

where C_o and C_e are initial and equilibrium concentrations of MO in mg/L, W is the mass of adsorbent in g, and V is the volume of solution in mL.

3. Result and discussion

3.1 Characterization of materials

The effect of the addition of MCM-41 on characteristic of the obtained MCM-41/ZIF-67 composites has been determined from the results of characterization using XRD, FTIR, SEM-EDX, nitrogen adsorption-desorption, and TGA. The XRD patterns of ZIF-67 and MCM-41/ZIF-67 composites are shown in **Figure 1**. As can be seen, the characteristic peaks appear at 2θ of 7.32, 10.36, 12.69, 14.66, and 16.40°,

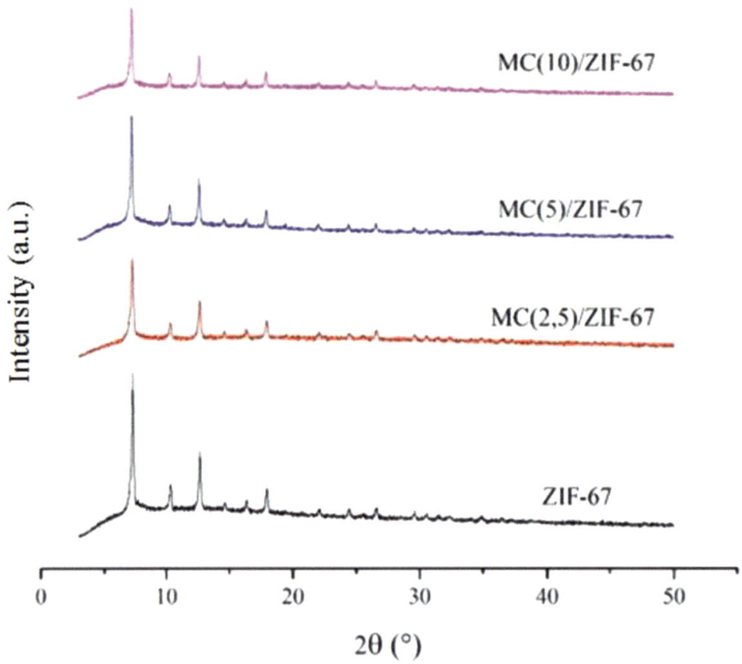

Figure 1.
XRD pattern of ZIF-67 and MCM-41/ZIF-67 composites.

respectively, which are in accordance with the results of previous studies [22]. Likewise, the XRD patterns of the MCM-41/ZIF-67 composites have shown the same pattern as pure ZIF-67. The addition of MCM-41 during the ZIF-67 synthesis, however, has resulted in a reduction in crystallinity of the obtained composites, which may be attributed to amorphous properties of mesoporous silica [24].

FT-IR analysis has been carried out to identify the functional groups of the obtained materials, and the results are shown in **Figure 2**. Each sample exhibit the presence of a sharp absorption band at a wavenumber of 423 cm^{-1}, due to the Co—N stretching vibration, indicating a bond between Co and N of the 2-methylimidazole ligand [25]. An absorption band at 693 cm^{-1} may be attributed to the bending vibration of the ring 2-metilimidazole [26]. Absorption peaks at wavenumbers of 755 and 1577 cm^{-1} may be resulted from out-of-plane vibrations and stretching vibrations of C = N in 2-methylimidazole ligands. The occurrence of C—N bending vibration has been identified from the appearance of a peak at wavenumber of 992 cm^{-1}, while stretching vibration from C—N is identified at wavenumber 1141 cm^{-1}. Absorption bands at wavenumbers 2926 and 3134 cm^{-1} appear due to stretching vibration of C—H sp^3 aromatic ring on 2-methylimidazole and stretching vibration of C—H sp^2 on aliphatic hydrocarbon chains [25]. The characteristic peak of MCM-41 that appears at wavenumber of 1050 cm^{-1}, as a result of Si—O—Si asymmetric vibration, is seen in the MC(5)/ZIF-67 composite spectrum and appears more pronounced in the MC(10)/ZIF-67 composite. This observation has indicated the existence of MCM-41 in the composite.

Surface morphology of ZIF-67, as is shown in **Figure 3(a)**, has a cubic shape, similar to that reported by Qian et al. [18]. Likewise, the MC(10)/ZIF-67 composite also has a uniform surface morphology, as is shown in **Figure 3(b)**. The results of SEM showed that the presence of mesoporous silica resulted in the increase in particle size, while the cube shape and more uniform surface morphology remained unchanged. The SEM images may indicate that the growth of ZIF-67 has taken place

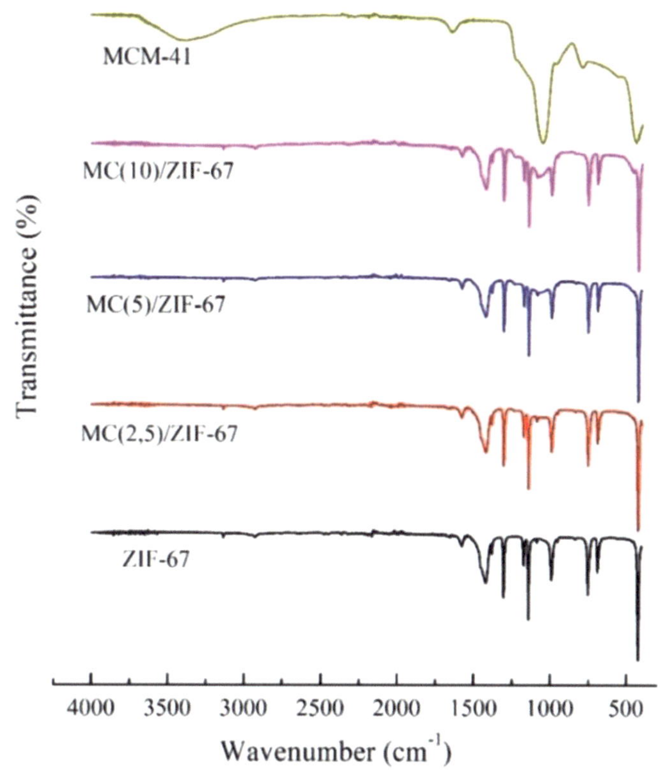

Figure 2.
FT-IR spectra of ZIF-67 and the composites.

inside the pores as well as on the surface of mesoporous silica. The phenomenon is reinforced by TEM image of the MC(10)/ZIF-67 composite, in which the characteristic hexagonal pore shape of MCM-41 is not seen any more in **Figure 3(c)**.

Nitrogen adsorption-desorption isotherms and pore size distributions of ZIF-8, and the composites are presented in **Figure 4(a)** and **(b)**, respectively, and the results of BET surface area and pore size are summarized in **Table 1**.

As shown in **Figure 4(a)**, type I isotherm adsorption curves of ZIF-67 and the composites have indicated that adsorption occurred in micropores of the samples that have high surface area. At a relatively higher pressure (P/P_0 = 0.85–1), the amount of gas absorbed increased and a narrow vertical hysteresis loop was observed on all samples, which may be due to the presence of interparticle mesoporous. Based on the BET equation, the highest surface area was observed for

(a) (b) (c)

Figure 3.
SEM image of (a) ZIF-67 (b) MC (10)/ZIF-67 (c) and TEM image of MC (10)/ZIF-67.

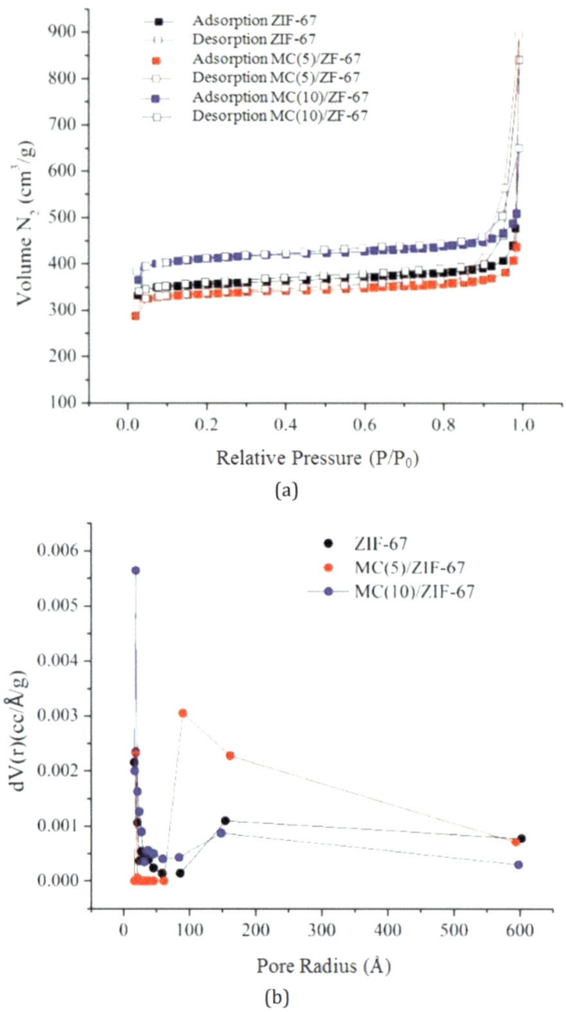

Figure 4.
(a) N_2 adsorption-desorption isotherm and (b) pore distribution of ZIF-67 and the composites.

MC(10)/ZIF-67 (1250.6 m^2/g). The high surface area may result from the formation of ZIF-8 particles inside the MCM-41 mesopores so as to produce new micropores with high surface area [27]. The pore distribution curves of ZIF-67 and the composites based on the BJH method are shown in **Figure 4b**, in which microporous and mesoporous distributions have been observed (**Table 1**). It can be concluded that the addition of mesoporous silica into the ZIF-8 synthesis results in the increase in the pore volume and pore diameter of the obtained composites (**Figure 4b**).

The results of TGA, as is shown in **Figure 5**, confirmed two-stage process of decomposition of ZIF-67 and MC(10)/ZIF-67 composite.

At a temperature range of 35–325°C, a mass decrease of 7.68% was observed, as a result of the loss of guest molecules of solvent or ligands that occupied the pores or surface of the samples [18]. Furthermore, a sharp decrease of 60.05% mass was observed in the temperature range of 325–429°C, which considered as decomposition of 2-methylimidazole ligand and damage of ZIF-67 structure. In addition, it can be seen in **Figure 5** that the thermal stability of the MC(10)/ZIF-67 was slightly higher than that of ZIF-67. The observation was in accordance with the result reported by Bhatt et al. [28]. Also, higher remaining mass of the composite as compared to the ZIF-67 was due to presence of SiO_2 from the MCM-41 added.

Materials	S_{BET}[a] (m^2/g)	V_{meso}[b] (cm^3/g)	V_{micro}[c] (cm^3/g)	D[d] (nm)
MCM-41	887.5	0.33	0.33	3.81
ZIF-67	1079.2	0.83	0.47	4.82
MC(5)/ZIF-67	1011.2	0.98	0.39	5.47
MC(10)/ZIF-67	1250.6	0.50	0.50	3.21

[a]*BET surface area.*
[b]*Mesopore volume calculated using BJH method.*
[c]*Micropore volume calculated using t-plot method.*
[d]*Average pore diameter.*

Table 1.
Summary of the textural properties of the synthesized material at different addition of MCM-41.

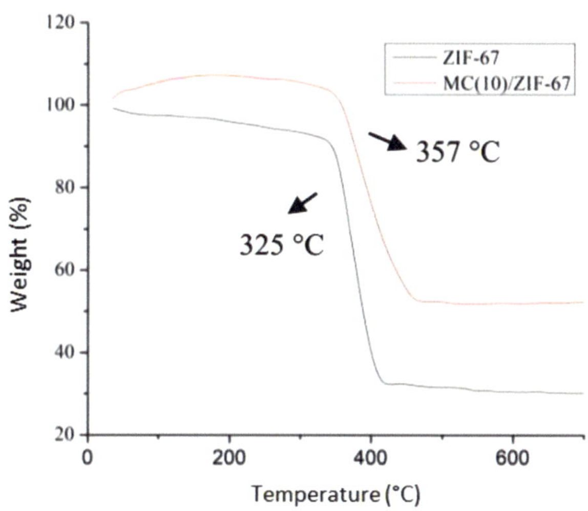

Figure 5.
Thermogravimetric profiles of ZIF-7 and MC(10)/ZIF-67 composite.

3.2 Adsorption kinetic studies

Kinetic study of MO adsorption was carried out using 20 mL of MO solution with a concentration of 100 mg/L. The contact times were varied in the range of 5–40 min, while the weight of adsorbent was 10 mg. The effect of contact time on MO adsorption for ZIF-67, MC(5)/ZIF-67, and MC(10)/ZIF-67 materials is illustrated in **Figure 6**. It can be seen that adsorption capacities increased with the increase in contact time to reach a maximum value after 25 min, after which a more stable adsorption capacity was observed until 50 min. The results indicated that the empty space of pores and the surface of adsorbent have been completely filled with adsorbate [29].

The kinetics data were analyzed using pseudofirst-order and pseudosecond-order kinetic models using Eqs. (3) and (4).

$$Ln(Q_e - Q_t) = Ln\,Q_e - K_1 t \tag{3}$$

$$\frac{t}{Q_t} = \frac{1}{K_2 Q_e^2} + \frac{t}{Q_e} \tag{4}$$

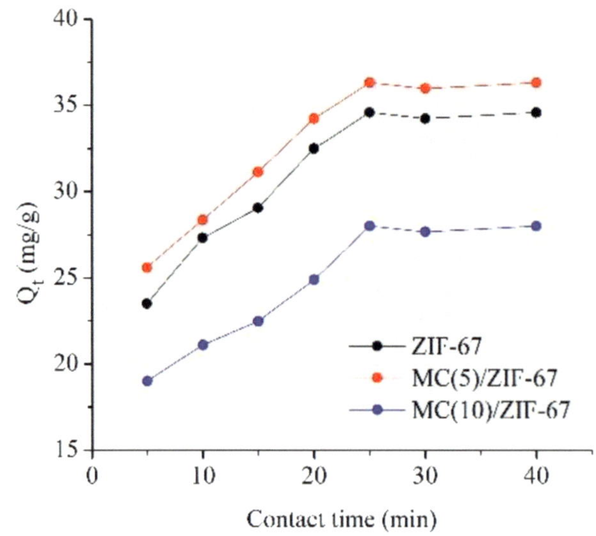

Figure 6.
Effect of contact time on MO adsorption.

Based on the calculation of adsorption kinetics with Eqs. (3) and (4), graph plots were performed on each of the adsorption kinetics models and the obtained graphs are shown in **Figure 7**. MO adsorption kinetics model of ZIF-67, MC(5)/ZIF-67, and MC(10)/ZIF-67 adsorbents were determined by comparing the R^2 values of each graph of the adsorption kinetics as shown in **Table 2**. It can be seen that correlation coefficient of the pseudosecond-order plot is higher than the pseudofirst-order plot. The results indicated that the mechanism of adsorption in all materials observed was as a result of chemisorption [30].

Adsorption of MO at different initial concentrations was carried out at the optimum contact time obtained of 25 min with MO concentration in the range of 50–330 mg/L. As is shown in **Figure 8**, the adsorption capacity increased with the increase in initial concentration of MO and the equilibrium of adsorption capacity was observed at an initial concentration of 285.20 mg/L. By further increasing the initial MO concentrations, a stable equilibrium adsorption capacity was obtained.

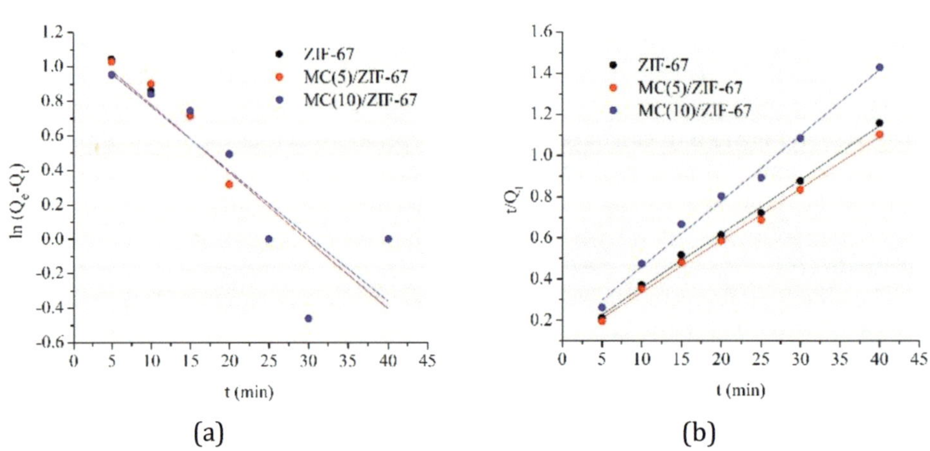

Figure 7.
Plots of (a) pseudofirst-order kinetics and (b) pseudosecond-order kinetics for the adsorption of MO.

Adsorbent	$Qe_{(exp)}$	Pseudofirst order			Pseudosecond order		
		$Qe_{(cal)}$	K_1	R^2	$Qe_{(cal)}$	K_2	R^2
ZIF-67	34.5782	3.2381	0.0395	0.7534	38.0228	0.00737	0.9967
MCM-41/ ZIF-67(5)	36.3071	3.2407	0.0395	0.7545	39.6825	0.00752	0.9970
MCM-41/ ZIF-67(10)	28.0083	3.1525	0.0377	0.7377	31.3480	0.00726	0.9923

Table 2.
Kinetic parameters and correlation coefficients for the pseudofirst-order and pseudosecond-order equations.

The MC(5)/ZIF-67 showed highest Q_e value of 167.64 mg/g, compared to other adsorbents. The results of adsorption studies have been in accordance with the results of N_2 adsorption-desorption analysis, in which the higher the pore size of the adsorbent, the higher is the adsorption capacity. The largest pore size of MC(5)/ZIF-67, as a result of adding 5%w/w mesoporous silica during the ZIF-67 synthesis, has improved its performance in MO adsorption [26]. The large pore size of the MC(5)/ZIF-67 allowed MO molecules to easily diffuse in to the pores and interacted electrostatically with cobalt ions from ZIF-67.

3.3 Isotherm adsorption

The study of adsorption isotherms was based on Langmuir and Freundlich isotherms models as calculated in Eqs. (5) and (6):

$$\frac{1}{Q_e} = \frac{1}{Q_m} + \frac{1}{Q_m \times K_L \times C_e} \quad (5)$$

$$LnQ_e = LnK_F + \frac{1}{n}LnC_e \quad (6)$$

Plots of adsorption isotherm obtained from the calculation are presented in **Figure 9**. The calculated correlation coefficient for each adsorbent used approaches

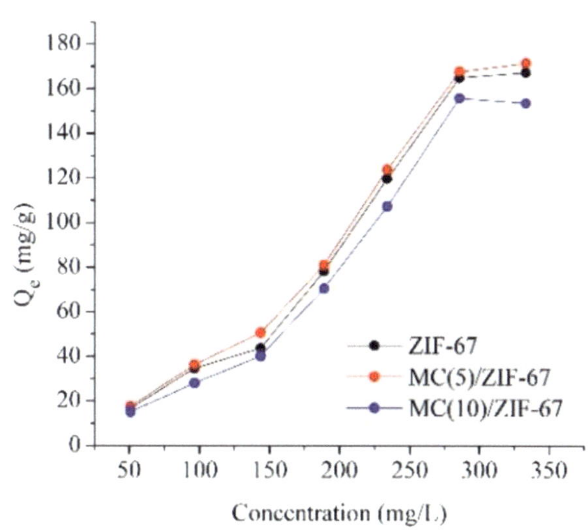

Figure 8.
Effect of initial concentrations of MO.

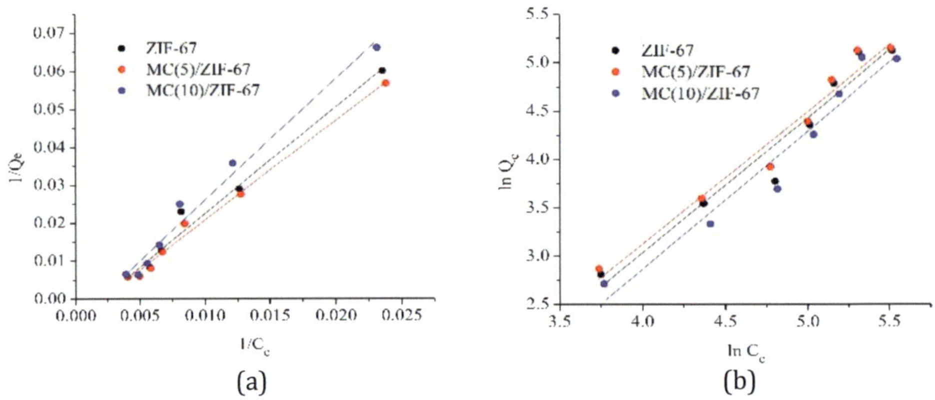

Figure 9.
Plots of (a) Langmuir and (b) Freundlich models of MO adsorption.

Adsorbent	$Qe_{(exp)}$	Langmuir			Freundlich		
		Q_m (mg/g)	K_L (L/mg)	R^2	K_F (mg/g)	n	R^2
ZIF-67	164.877	175.439	0.0020	0.982	12.705	0.7172	0.946
MC(5)/ZIF-67	167.635	181.818	0.0021	0.992	10.578	0.7296	0.965
MC(10)/ZIF-67	155.878	166.667	0.0019	0.981	17.0134	0.7018	0.956

Table 3.
Langmuir and Freundlich isotherm parameters for the adsorption of MO.

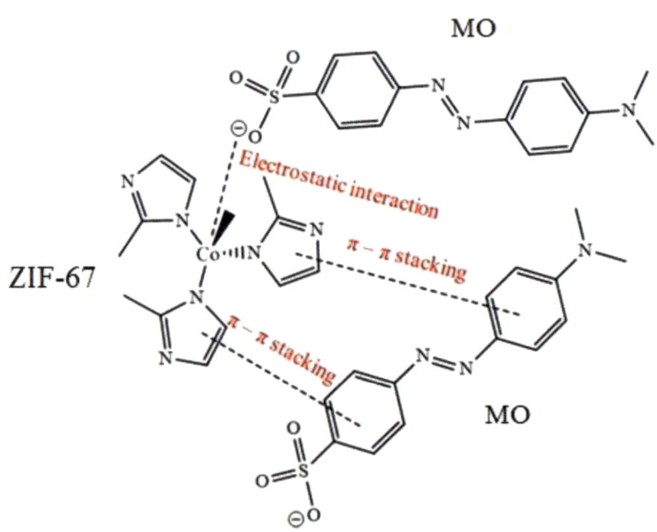

Figure 10.
Mechanism of MO adsorption to ZIF-67.

with the value of 1 in the Langmuir model (**Table 3**). These results indicated that the adsorption process occurred in accordance with the Langmuir adsorption isotherm model, which led to the assumption that adsorption occurred on a specific homogeneous surface of the adsorbent, which had the same energy for monolayer adsorption [31].

As an anionic dye, MO contains SO_3 group in its molecular structure. The adsorption of MO by ZIF-67 and MCM-41/ZIF-67 composites can occur as a result of the electrostatic interaction between the SO_3 group and the positive charge of cobalt ion from ZIF-67 [12]. In addition, the imidazole rings of ZIF-67 may interact with the benzene rings from MO molecules through π-π stacking [16, 19]. The molecular interaction of MO with ZIF-67 is illustrated in **Figure 10**.

4. Conclusions

ZIF-67 and MCM-41/ZIF-67 composites have been successfully synthesized at room temperature in aqueous solution with addition of TEA as a deprotonating agent. The results of characterization using XRD indicated that both ZIF-67 and composites exhibited the same diffractogram patterns, with characteristic peaks at 2θ of 7.32, 10.36, 12.69, 14.66, and 16.40°, similar to that of reported ZIF-67. In addition, the absorption bands of FTIR spectra of all composites appeared at the same wavenumber as observed for synthesized and the reported ZIF-67. SEM images showed similar cube shapes of surface morphology between ZIF-67 and MC(10)/ZIF-67 composites. The results of analysis by adsorption-desorption N_2 showed that the highest surface area of 1250.6 m^2/g was observed for MC(10)/ZIF-67. The highest adsorption capacity toward MO in aqueous solution of 167.6 mg/g was achieved by MC(5)/ZIF-67 adsorbent, and followed a pseudosecond-order adsorption kinetics and Langmuir adsorption isotherm.

Acknowledgements

We gratefully acknowledge funding from Institut Teknologi Sepuluh Nopember (ITS) Surabaya, through Laboratory Research Grant 2018, and Material Chemistry and Energy Laboratory, Department of Chemistry-ITS, for facilities supporting this work.

Author details

Ratna Ediati*, Pramita Elfianuar, Eko Santoso, Dety Oktavia Sulistiono and Muhammad Nadjib
Department of Chemistry, Institut Teknologi Sepuluh Nopember (ITS), Surabaya, Indonesia

*Address all correspondence to: ratna.ediati@gmail.com

References

[1] Chatterjee S, Lee DS, Lee MW, Woo SH. Congo red adsorption from aqueous solutions by using chitosan hydrogel beads impregnated with nonionic or anionic surfactant. Bioresource Technology. 2009;**100**:3862-3868. DOI: 10.1016/j.biortech.2009.03.023

[2] Shi L, Hu L, Zheng J, Zhang M, Xu J. Adsorptive removal of methylene blue from aqueous solution using a Ni-metal organic framework material. Journal of Dispersion Science and Technology. 2016;**37**:1226-1231. DOI: 10.1080/01932691.2015.1050731

[3] Chen S, Huang Y, Han X, Wu Z, Lai C, Wang J, et al. Simultaneous and efficient removal of Cr(VI) and methyl orange on LDHs decorated porous carbons. Chemical Engineering Journal. 2018. DOI: 10.1016/j.cej.2018.07.012

[4] Habiba AU, Siddique TA, Jia J, Lee L, Joo TC, Ang BC, et al. Adsorption study of methyl orange by chitosan/polyvinyl alcohol/zeolite electrospun composites nanofibrous membrane. Carbohydrate Polymers. 2018. DOI: 10.1016/j.carbpol.2018.02.081

[5] Gracia F, Lee AF, Mansoob M. Mechanothermal synthesis of Ag/TiO_2 for photocatalytic methyl orange degradation and hydrogen production. Process Safety and Environmental Protection. 2018. DOI: 10.1016/j.psep.2018.09.015

[6] Khan NA, Hasan Z, Jhung SH. Adsorptive removal of hazardous materials using metal-organic frameworks (MOFs): A review. Journal of Hazardous Materials. 2012:1-13. DOI: 10.1016/j.jhazmat.2012.11.011

[7] Houas A et al. Photocatalytic degradation pathway of methylene blue in water. Applied Catalysis B: Environmental. 2001;**31**:145-157. DOI: 10.1016/S0926-3373(00)00276-9

[8] Ayed L, Mahdhi A, Cheref A, Bakhrouf A. Decolorization and degradation of azo dye Methyl Red by an isolated *Sphingomonas paucimobilis*: Biotoxicity and metabolites characterization. Desalination. 2011;**274**:272-277. DOI: 10.1016/j.desal.2011.02.024

[9] Li SH, Zhao Y, Chu J, Li WW, Yu HQ, Liu G. Electrochemical degradation of methyl orange on Pt-Bi/C nanostructured electrode by a square-wave potential method. Electrochimica Acta. 2013;**92**:93-101. DOI: 10.1016/j.electacta.2013.01.012

[10] Bamperng S, Suwannachart T, Atchariyawut S, Jiraratananon R. Ozonation of dye wastewater by membrane contactor using PVDF and PTFE membranes. Separation and Purification Technology. 2010;**72**:186-193. DOI: 10.1016/j.seppur.2010.02.006

[11] Yang Y, Yan X, Hu X, Feng R, Zhou M. In-situ growth of ZIF-8 on layered double hydroxide: Effect of Zn/Al molar ratios on their structural, morphological and adsorption properties. Journal of Colloid and Interface Science. 2017;**505**:206-212. DOI: 10.1016/j.jcis.2017.05.100

[12] Zhang ZH, Zhang JL, Liu JM, Xiong ZH, Chen X. Selective and competitive adsorption of azo dyes on the metal–organic framework ZIF-67. Water, Air, and Soil Pollution. 2016;**227**. DOI: 10.1007/s11270-016-3166-7

[13] Khan NA, Jung BK, Hasan Z, Jhung SH. Adsorption and removal of phthalic acid and diethyl phthalate from water with zeolitic imidazolate and metal-organic frameworks. Journal of Hazardous Materials. 2015;**282**:194-200. DOI: 10.1016/j.jhazmat.2014.03.047

[14] Tan K, Zuluaga S, Gong Q, Canepa P, Wang H, Li J, et al. Water

reaction mechanism in metal organic frameworks with coordinatively unsaturated metal ions: MOF-74. Chemistry of Materials. 2014;**26**: 6886-6895. DOI: 10.1021/cm5038183

[15] Cho HY, Yang DA, Kim J, Jeong SY, Ahn WS. CO_2 adsorption and catalytic application of Co-MOF-74 synthesized by microwave heating. Catalysis Today. 2012;**185**:35-40. DOI: 10.1016/j.cattod.2011.08.019

[16] Hasan Z, Jhung SH. Removal of hazardous organics from water using metal-organic frameworks (MOFs): Plausible mechanisms for selective adsorptions. Journal of Hazardous Materials. 2015;**283**:329-339. DOI: 10.1016/j.jhazmat.2014.09.046

[17] Park KS, Ni Z, Côté AP, Choi JY, Huang R, Uribe-Romo FJ, et al. Exceptional chemical and thermal stability of zeolitic imidazolate frameworks. Proceedings of the National Academy of Sciences of the United States of America. 2006;**103**:10186-10191. DOI: 10.1073/pnas.0602439103

[18] Qian J, Sun F, Qin L. Hydrothermal synthesis of zeolitic imidazolate framework-67 (ZIF-67) nanocrystals. Materials Letters. 2012;**82**:220-223. DOI: 10.1016/j.matlet.2012.05.077

[19] Lin KA, Chang H. Chemosphere ultra-high adsorption capacity of zeolitic imidazole framework-67 (ZIF-67) for removal of malachite green from water. Chemosphere. 2015;**67**:1-8. DOI: 10.1016/j.chemosphere.2015.01.041

[20] Kondo A, Takanashi S, Maeda K. New insight into mesoporous silica for nano metal-organic framework. Journal of Colloid and Interface Science. 2012. DOI: 10.1016/j.jcis.2012.06.040

[21] Zhao XS, Lu GQ. Modification of MCM-41 by surface silylation with trimethylchlorosilane and adsorption study. The Journal of Physical Chemistry. B. 1998. DOI: 10.1021/jp972788m

[22] Gross AF, Sherman E, Vajo JJ. Aqueous room temperature synthesis of cobalt and zinc sodalite zeolitic imidizolate frameworks. Dalton Transactions. 2012;**41**:5458. DOI: 10.1039/c2dt30174a

[23] Badamali S, Sakthivel A, Selvam P. Influence of aluminium sources on the synthesis and catalytic activity of mesoporous AlMCM-41 molecular sieves. Catalysis Today. 2000;**63**:291-295. DOI: 10.1016/S0920-5861(00)00471-5

[24] Chen Y, Zhang X, Dong M, Wu Y, Zheng G, Huang J, et al. MCM-41 immobilized 12-silicotungstic acid mesoporous materials: Structural and catalytic properties for esterification of levulinic acid and oleic acid. Journal of the Taiwan Institute of Chemical Engineers. 2016;**61**:147-155. DOI: 10.1016/j.jtice.2015.12.005

[25] Verpoort F. Materials Chemistry A. 2016. DOI: 10.1039/C6TA07860E

[26] Hartman BT, Brand R. Thesis for the degree of Master of Science In the Graduate School of Natural Sciences Debye Institute for Nanomaterials Science Inorganic Chemistry and Catalysis, (n.d.)

[27] Tari NE, Tadjarodi A, Tamnanloo J, Fatemi S. One pot microwave synthesis of MCM-41/Cu based MOF composite with improved CO_2 adsorption and selectivity. Microporous and Mesoporous Materials. 2016;**231**:154-162. DOI: 10.1016/j.micromeso.2016.05.027

[28] Bhatt AP, Pathak K, Jasra RV, Kureshy RI, Khan NUH, Abdi SHR. Chiral lanthanum-lithium-binaphthol complex covalently bonded to silica and MCM-41 for enantioselective nitroaldol (Henry) reaction. Journal of Molecular

Catalysis A: Chemical. 2006;**244**:110-117. DOI: 10.1016/j.molcata.2005.08.045

[29] Liu X, Luo J, Zhu Y, Yang Y, Yang S. Removal of methylene blue from aqueous solutions by an adsorbent based on metal-organic framework and polyoxometalate. Journal of Alloys and Compounds. 2015. DOI: 10.1016/j.jallcom.2015.07.065

[30] Lin S, Song Z, Che G, Ren A, Li P, Liu C, et al. Adsorption behavior of metal-organic frameworks for methylene blue from aqueous solution. Microporous and Mesoporous Materials. 2014. DOI: 10.1016/j.micromeso.2014.03.004

[31] Li L, Liu XL, Geng HY, Hu B, Song GW, Xu ZS. A MOF/graphite oxide hybrid (MOF: HKUST-1) material for the adsorption of methylene blue from aqueous solution. Journal of Materials Chemistry A. 2013. DOI: 10.1039/c3ta11478c

Chapter 4

Synthesis, Properties, and Their Potential Application of Covalent Organic Frameworks (COFs)

Lifeng Deng, Junfeng Zhang and Yanan Gao

Abstract

Covalent organic frameworks (COFs) represent an emerging class of crystalline porous polymers, which are ingeniously assembled with organic building blocks through reversible covalent bonds. The well-defined crystalline porous structures, easy functional modification, high surface area, together with structural designability and diversity have offered the COFs superior potential in various applications, such as catalysis, gas adsorption and separation, and optoelectricity. Since the pioneer work of Omar Yaghi in 2005, this field has developed rapidly and attracted much attention from researchers with diverse expertise. In this chapter, we describe the basic design concepts, the recent synthetic advancements, and the frontiers of functional exploration in gas adsorption and storage. Special emphasis is placed on their potential application in heterogeneous catalysis field. Finally, the prospects of COFs and remaining issues in these fields are indicated.

Keywords: covalent organic framework, porous organic polymers, catalysis, gas adsorption, topological structure

1. Introduction

Over the years, interest in the field of porous materials has grown extremely due to their good performance and potential applications in gas storage and separation, catalysis, energy conversion and storage, and optoelectronics. Although a variety of porous materials have been synthesized by various ways, it has proven difficult to prepare organic polymer networks with discrete pores. However, the appearance of reticular chemistry makes it possible to construct these porous materials, which can be synthesized by using topologically designed building blocks.

The first family of such porous materials synthesized based on reticular chemistry is metal-organic frameworks (MOFs), which consist of metal ions or clusters coordinated to organic ligands to form one-, two-, or three-dimensional network structures [1]. The coordination versatility of the metal ions, combined with the functional diversity of the organic linkers gives these materials immense possibilities. Another family of porous materials to be synthesized on the base of reticular chemistry is covalent organic frameworks (COFs), which were firstly demonstrated by Yaghi and co-workers in 2005 [2]. Different from MOFs, COFs are constructed by pure organic building blocks to form crystalline porous polymers via reversible covalent bonds with predictable control over composition, topology, and porosity [3, 4]. Therefore, COFs could be regarded as "organic zeolites."

In principle, the formation of COFs by linking organic building blocks is expected to be thermodynamically unfavorable. In many cases, amorphous cross-linked polymers or poorly defined solids are often obtained through the linkage of organic building blocks with strong covalent bonds. In order to obtain a crystalline material, the formation of linkages must be reversible and the reaction rates should be on a time scale that allows for defects to be self-corrected. Dynamic covalent chemistry (DCC) realizes the reversible formation of strong covalent bonds, which can be formed, broken, and reformed under certain conditions [5]. DCC is thermodynamically controlled and can endow reversible reaction systems with "error checking" and "proof-reading" features, which produces the most thermodynamically stable structures. Therefore, reversible covalent bond as linkages is considered to be prerequisite to construct crystalline COF materials. So far, the organic linkages involved in the formation of COFs include B-O (boroxine, boronate ester, spiroborate, and borosilicate), C=N (imine, hydrazone, and squaraine), C-N (β-ketoenamine, imide, and amide), etc. (**Figure 1**) [3].

In addition to reversible covalent bonds, the structure of the building blocks has to be considered to obtain crystalline COFs. During the formation process of COFs, the geometry of the building blocks needs to be well preserved in the COF framework. So, the desired building blocks should be structurally rigid, and the formation direction of reversible covalent bonds must be discrete [3]. From the

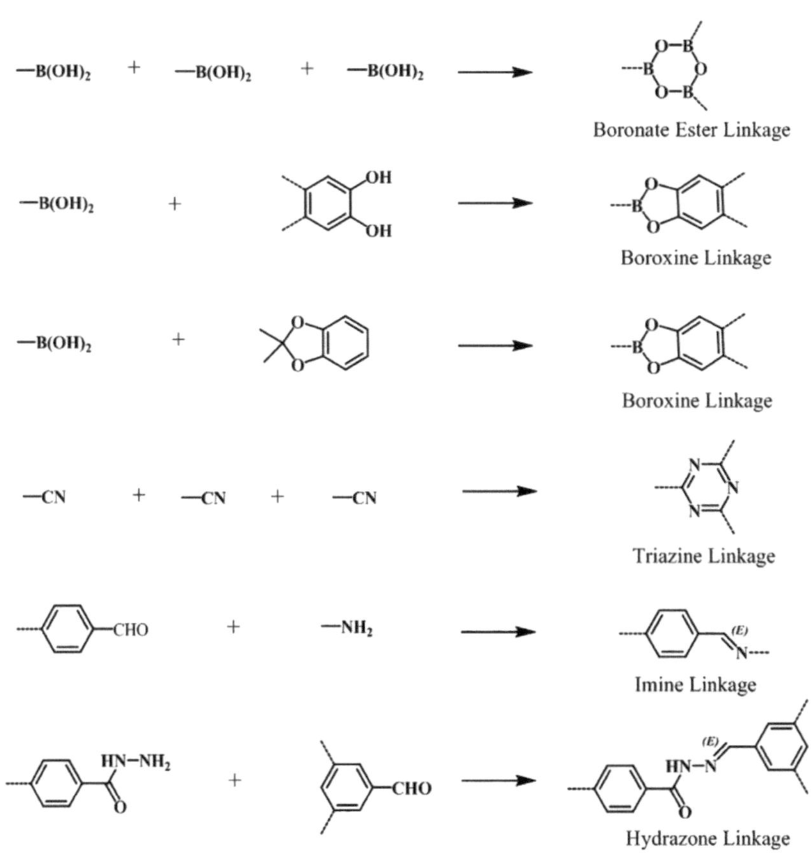

Figure 1.
Schematic representation of the dynamic chemical reactions for the preparation of various COFs ([3]-Reproduced by permission of The Royal Society of Chemistry).

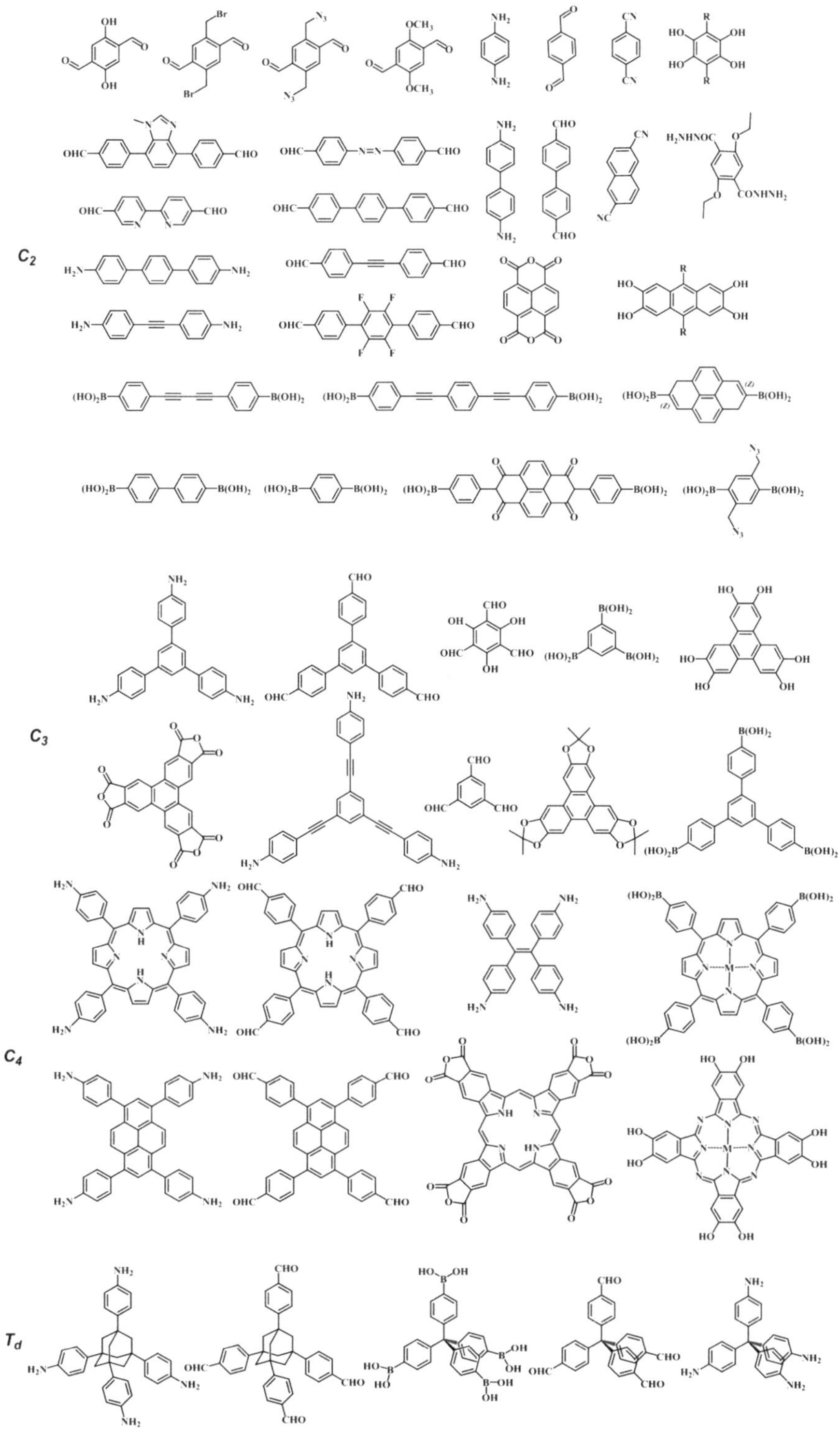

Figure 2.
The geometry of the building blocks for the preparation of COFs.

viewpoint of geometric structure, these building blocks can be classified into 2D or 3D structure based on simplified symmetry notation, which refers to the directional symmetry of the reactive groups. The reported building blocks (2D-C_2, 2D-C_3,

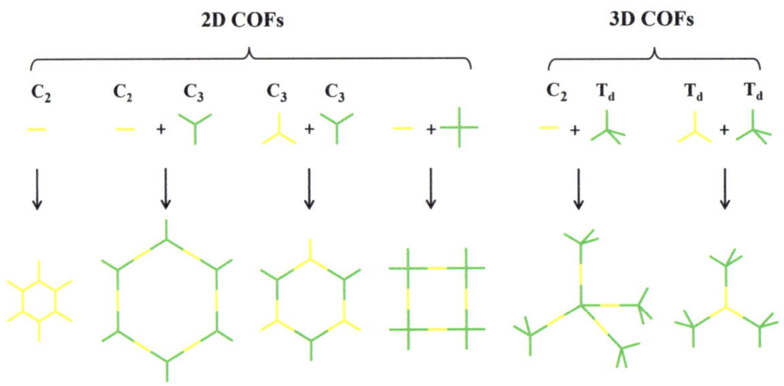

Figure 3.
The combination of building blocks with different geometries to design COFs.

2D-C_4, and 3D-T_d) are summarized in **Figure 2**. The COF structure is determined by the geometry of the building blocks. For example, the combinations of 2D building blocks, like 2D-C_2 + 2D-C_3, 2D-C_3 + 2D-C_3, or 2D-C_2 + 2D-C_4, will lead to 2D COFs with predesigned topology and pore structure (**Figure 3**).

In contrast, the self-condensation of T_d building blocks (3D-T_d + 3D-T_d) or the combinations of T_d building blocks with 2D building blocks such as 3D-T_d + 2D-C_2, 3D-T_d + 2D-C_3, or 3D-T_d + 2D-C_4 allow for the construction of 3D COFs with different crystalline networks (**Figure 3**). Unlike the rapid expansion of 2D COFs, the progress in 3D COF chemistry has been severely impeded due to the limited choice of monomers. Therefore, only a few 3D COFs have been reported so far [6]. The rigid structure and discrete bonding direction of aromatics make them ideal building blocks for constructing COFs. The diversity of aromatic compounds allows a number of building block combinations, which makes it possible that the COFs have high flexibility in their molecular design.

2. Synthesis of COFs and their properties

Though many building units have been developed, it is still not an easy issue for constructing COFs with both high porosity and crystallinity. Because, it is also important to find suitable conditions under which reversible bond formation is possible without resorting to extreme temperatures or pressures. In other words, a synthetic method is another key to obtain crystalline COFs. A suitable temperature and pressure are required, and the appropriate solvent combinations and ratios are of specially importance in getting the balance between framework formation and crystallization [3]. In general, a high temperature (80–120°C) combined with a low reaction pressure is used to synthesize COFs. As for reaction medium, to date, almost all COFs, as either bulk materials or thin films grown on the support surface, have been obtained under solvothermal synthesis conditions. A traditional organic solvent or a solvent mixture has been used as a reaction medium. Besides, molten metal salts, including molten $ZnCl_2$ and ionic liquids, have been attempted to provide ionothermal conditions for the synthesis of COFs. Regarding heating fashion to reaction system, microwave-assisted heating has proved to be quick to synthesize COFs under solvothermal conditions.

2.1 Solvothermal synthesis

As mentioned above, most of the reported COFs were obtained through the solvothermal synthesis method. The solvothermal synthesis of COFs often takes

3–7 days and requires heating at a high temperature within a sealed vessel. The pressure of the reaction system is of importance, which may affect the crystallinity and porosity of COFs significantly. In many cases, a 150 mTorr is the optimal pressure before the reaction system is sealed within a ~10 cm^3 vessel. In addition, the solvent chosen for the formation of COFs is crucial as it governs the solubility of reactants and is closely related to reaction rate, crystal nucleation, crystal growth rate, and self-healing structure. Therefore, how to find an appropriate solvent system to get crystalline COFs has long been an intractable problem. A binary solvent mixture was generally used to obtain COFs in the presence of an acidic catalyst. For instance, solvent mixtures such as dioxane-mesitylene [2], DMAc-o-dichlorobenzene [7], and THF-methanol [8] have been employed to construct boronate ester-linked COFs. A mixture of dioxane-ethanol has been used as the solvent to synthesize imine-linked COFs [9], and even a ternary solvent system has been used to synthesize COF materials, in which a mixture of N-methyl -2-pyrrolidone (NMP), mesitylene, and isoquinoline afforded crystalline PI-COF-4 and PI-COF-5 with high yields after 5 days reaction [10]. In some cases, a single solvent has been found to be effective for getting crystalline COFs. For instance, a 3D COF, COF-DL229, has been recently synthesized by our group through the condensation reaction of tetrahedral 1,3,5,7-tetrakis(4-aminophenyl)-adamantane (TAPA) and linear 1,4-phthalaldehyde (PTA) in a sole solvent, 1,2-dichlorobenzene, and the resultant 3D COF exhibits high crystallinity and high BET surface area (1762 m^2 g^{-1}) [11]. In addition, a suitable concentration of building blocks or the content of catalysis used in the solvent system is of significance for the reaction to proceed under thermodynamic control [7, 8, 12].

2.2 Ionothermal synthesis

Thomas and co-workers reported an ionothermal synthesis strategy to produce COFs [13, 14]. The cyclotrimerization of aromatic nitrile building units in molten $ZnCl_2$ at 400°C affords porous covalent triazine-based frameworks (CTFs), a special class of COFs, with robust chemical and thermal stabilities. Molten $ZnCl_2$ acts as both the solvent and the catalyst for the reversible cyclotrimerization reaction. Compared with those COFs obtained under solvothermal synthesis method, the CTFs have the disadvantage in crystallinity control because the reversible cyclotrimerization reaction takes place under very harsh reaction conditions. Also, the harsh reaction conditions limit the building block availability. Most synthesized CTFs are amorphous structures that lack long-range molecular order. To date, only two building units have been successfully used to synthesize the crystalline CTFs [13, 14].

Recently, a common ionic liquid, 1-butyl-3-methylimidazolium bis((trifluoromethyl)sulfonyl)imide ([bmim][Tf_2N]) has been used as solvent to prepare highly crystalline COF materials [15]. It is well known that ionic liquids are a class of low-temperature molten salts (with molten point less than 100°C) and they are widely regarded as environmentally benign media because they can be used as an alternative to the traditional organic solvents [16]. Ionic liquids are nonvolatile, nonflammable, and their properties can be changed by counter anions. These interesting properties make them very attractive in many important reactions in industrial field. Particular interest has been shown in the preparation of molecular sieves [17] and metal-organic frameworks (MOFs) [18, 19]. Since ionic liquids have a neglectable vapor pressure, this strategy displays significant merits by eliminating any safety concerns associated with high autogenous pressures. Besides, some novel architectures have often been obtained when ionic liquids are used instead [17]. In this regard, Qiu and his co-workers firstly reported the synthesis of several 3D

COFs by using ionic liquids as reaction medium [15]. Crystalline 3D COFs can be obtained under very mild conditions. A high-reaction speed has been observed, for example, 3D-IL-COF-1 can be obtained within only 3 minutes. Furthermore, it was found that the ionic liquid can be reused without obvious activity loss. This work not only provides a new strategy to synthesize COFs but also opens a potential way to green large-scale COF synthesis in industry.

2.3 Microwave synthesis

Microwave synthesis is to apply microwave radiation in chemical reactions. Microwave heating is an effective way to accelerate the chemical reactions. This technology has been widely used to synthesize crystalline MOF materials [20]. Microwave heating can possess certain benefits over conventional oven heating, including reaction rate acceleration, milder reaction conditions, higher product yield, lower energy usage, different reaction selectivities, and so on. Therefore, a rapid microwave-assisted method has been reported by Cooper and his co-workers for the synthesis of COF materials [21, 22]. Using microwave heating, both 2D COF-5 and 3D COF-102 can be obtained in 20 minutes, which is much faster than the reaction time of 72 hours required under solvothermal synthesis condition [2, 23]. Moreover, microwave heating affords a BET surface area of 2019 $m^2\ g^{-1}$ for COF-5, which is slightly higher than that synthesized solvothermally (1590 $m^2\ g^{-1}$).

2.4 Properties of COFs

The most important concern of porous materials is porosity. For traditional crystalline porous materials, like zeolite, structure-directing agents are generally used to template the formation of porous structure. The formation mechanism of COFs has been rarely reported, but it is generally accepted that the building blocks are first condensed into oligomers, which is similar to the step-growth of crystallization. Then, a nucleation process may afford metastable polymers or templated polymers. The small oligomers were further grown through the persistent linkage and stacking of building blocks, and simultaneously, the defects of amorphous polymers were repaired by a broken and reformed process, until the crystals achieve the bulk ascertained by X-ray diffraction, affording an extended crystalline porous structure with high surface area [24]. The molecular structure of building blocks governs the porosity of COFs. The length of building blocks determines the pore size of COFs, and the geometrical shape determines the topological structures of COFs, which has been shown in **Figure 3**. The designability of COFs in porosity is a distinct difference from inorganic zeolites. The COFs with porosity in different sizes and shapes have been widely reported and will not be discussed here.

The surface areas of COFs are comparable to those of MOFs, but are generally larger than those of zeolites. The earliest reported COF-1 and COF-5 exhibited a surface area of S_{BET} = 711 $m^2\ g^{-1}$; the pore volume V_p = 0.32 $cm^3\ g^{-1}$ (**gra**, $P6_3/mmc$) for COF-1 and the surface area of S_{BET} = 1590 $m^2\ g^{-1}$; and the pore volume V_p = 0.998 $cm^3\ g^{-1}$ (**bnn**, P6/mmc) for COF-5 [2]. The subsequently reported 2D COF-6, COF-8, and COF-10 demonstrated the surface areas of S_{BET} = 1049, 968, and 976 $m^2\ g^{-1}$ and the pore volume V_p = 0.32, 0.69, and 1.44 $cm^3\ g^{-1}$, respectively [25]. Dichtel et al. [26] used different sized building blocks to construct COFs. The condensations of Zn phthalocyanine (ZnPc) with pyrene diboronic acid, diphenyl butadiynediboronic acid, napthalenediimide diboronic acid, and PPE diboronic acid afford COFs with pore sizes of 2.7, 3.4, 4.0, and 4.4 nm, respectively. For 2D COFs, the surface areas of COFs are generally in the range from several hundreds to over 1000 m^2 per gram. However, a 2D mesoporous imine-based ILCOF-1 exhibited high surface area of S_{BET} = 2723 m^2 g^{-1} and S_{Lang} = 3453 m^2 g^{-1}, respectively [27]. This is the largest value reported to date for 2D COFs. In comparison with 2D COFs, 3D COFs possess relatively higher surface areas. Noninterpenetrated COF-105 (**ctn**) and COF-108 (**bor**) have surface areas of 3472 and 4210 m^2 g^{-1}, respectively [23]. The interpenetrated structures are inclined to reduce the surface area of in 3D COF nets. COF-300 with four- or fivefold interpenetration (**dia**) affords a surface area of S_{BET} = 1360 m^2 g^{-1} [28], and COF-320 with ninefold interpenetration **dia** net gives a surface area of S_{Lang} = 2400 m^2 g^{-1} [29], which is obviously lower than those of noninterpenetrated COF-105 and COF-108. Recently, a 3D COF with a record high surface area was reported by McGrier's group [30]. The linkage of π-electron-conjugated dehydrobenzoannulene and tetrahedraltetra-(4-dihydroxy-borylphenyl)methane produced 3D DBA-3D-COF 1. This COF achieves a surface area of S_{BET} = 5083 m^2 g^{-1}, a highest value among all COFs. Even though the COF was metalated with Ni, only a slight reduction in the surface area was observed and Ni-DBA-3D-COF 1 still has a surface area of S_{BET} = 4763 m^2 g^{-1}.

It seems that crystalline porous structure is inclined to provide higher surface area than amorphous materials. 3D COF-DL229, constructed by T_d-symmetric 1,3,5,7-tetrakis(4-aminophenyl)-adamantane and (TAPA) and 1,4-phthalaldehyde (PTA) by our group, shows a BET surface area of 1762 m^2 g^{-1} (Langmuir surface area of 1924 m^2 g^{-1}) [11]. Due to the organic component of COFs in nature, the skeleton of the COF-DL229 is "soft," and thus, the adsorption of vapor iodine within the pores of the COF led to the distortion of the skeleton. The XRD result indicated that the crystalline structure was destroyed, but the linkages were not broken by the captured iodine. The BET surface area of COF-DL229 was greatly decreased from 1762 to 87 m^2 g^{-1}, suggesting that the ordered crystalline structure of COFs afford a higher surface area than their corresponding amorphous counterparts.

3. Gas adsorption and storage

As a class of crystalline porous materials, COFs allow the atomically precise integration of pure organic units to create predesigned skeletons and nanopores based on reticular chemistry [3, 4]. Since the landmark work of Yaghi in 2005 [2], COFs have attracted extensive interest and have become one of the fastest growing fields in both material science and chemistry during the past decade. By far, a diversity of functional COF materials (either 2D or 3D structures) with tailored functionalities has been obtained. With tunable chemical and physical properties, these COF materials have been widely applied in various fields, such as in gas storage, photoelectricity, catalysis, etc.

COFs are composed of light elements linked by reversible covalent bonds to form porous materials with high surface areas. Therefore, COFs can be regarded as ideal materials for gas storage. The research of storage capabilities of COFs has mainly focused on hydrogen, methane, and carbon dioxide adsorption. Generally speaking, the gas adsorption performance of a COF material mainly depends on its components and topological structure. In comparison with 3D COFs, the dense π-stacking layer structure of 2D COFs gives rise to limited surface areas and small pore volumes [31], which restricts their potential for gas adsorption. By contrast, 3D COFs possess not only extremely high surface areas but also extraordinarily low densities, and furthermore, the entire material is accessible for adsorbed molecules to "sees" all the atoms of the framework [32]. Such impressive characteristics of 3D COFs make them promising candidates for adsorbent materials. As a result, 3D COFs possess significantly higher uptake capacities than 2D COFs. In this field, more theoretical simulation has been introduced to model the storage capacity of COFs [33, 34].

3.1 Hydrogen gas

Due to clean combustion and high chemical energy density, hydrogen has been pursued as an ideal substitute for traditional fossil fuels. The attempt to store hydrogen with COF materials has been recently reported. COF-18 Å (with a BET surface area of 1263 $m^2 g^{-1}$) shows a hydrogen uptake of 1.55% at 1 bar and 77 K [35]. However, COF-11 Å with a lower BET surface area of 105 $m^2 g^{-1}$ shows a hydrogen uptake of 1.22% under the same conditions. A 2D CTC-based COF, CTC-COF, with mesoporous structure, has been synthesized through solvothermal reaction. CTC-COF shows a surface area of 1710 $m^2 g^{-1}$ due to its bowl-shaped building block structure, which is larger than its analogue, COF-5 constructed by planar subunits. This material has a hydrogen uptake of 1.12 wt% at low pressure (1.05 bar) [36]. As expected, 3D COFs exhibited a better hydrogen uptake capability than 2D COFs. COF-102 shows a high hydrogen uptake of 7.24 wt% at saturation (35 bar, 77 K) [37], which is even higher than that of amorphous PAF-1 (7.0 wt%) [38]. NTU-COF-2, constructed by two kinds of reversible covalent binds, shows H_2 uptake as high as 174 $cm^3 g^{-1}$ (1.55 wt%) at 1.0 bar and 77 K, which makes it a top COF material for H_2 adsorption reported by far [39]. In 2012, Yaghi and his co-workers reported that a new COF-301-$PdCl_2$ reaches 60 g/L (100 bar), which is higher than the DOE 2015 target (40 g/L) and close to the ultimate target (2050) of 70 g/L [40]. At 1.0 bar, ILCOF-1 stores moderate amounts of hydrogen (1.3 wt.%, 77 K), though it has a high BET surface area of 2723 $m^2 g^{-1}$ [27]. Compared with experimental research, more theoretical work has been developed to evaluate the interaction of hydrogen gas with the COF pores [32, 41–44].

3.2 Methane gas

Methane is another important energetic gas. However, due to the lack of an effective, economic, and safe on-board storage medium, the development of methane-driven automobiles is greatly limited. 3D COF-102 shows a methane storage capacity of 187 mg g^{-1} (the target value for methane storage set by the DOE is 180 cm^3), and 3D COF-103 demonstrates a high adsorption capacity of 175 mg g^{-1} at 35 bar and 298 K [37], which are comparable to the highest observed values for MOFs (for instance, MOF-210; 220 mg g^{-1}) [45]. In contract, 2D COFs, including COF-1, COF-5, COF-6, COF-8, and COF-10, show methane storage capacity of 40, 89, 65, 87, and 80 mg g^{-1}, respectively, under the same conditions [37]. Similar to hydrogen storage, 3D COFs demonstrated higher storage capability than 2D COFs. By using a multiscale theoretical method, Cao and his coworkers calculated the storage capacities of methane in 3D COFs and their corresponding Li-doped COF materials at 243 and 298 K, respectively [46]. They predicted that at 298 K and 35 bar, COF-102 and COF-103 have excess gravimetric capacities of 17.72 and 16.61 wt%, respectively, which correspond to 302 and 285 cm^3 (STP)/g. It was shown that the predicted results are in agreement with the experimental ones. Goddard, III et al. designed 14 COFs expected to adsorb large amounts of methane at 298 K and up to 300 bar. They calculated the delivery uptake using grand canonical Monte Carlo (GCMC) simulations and found that among these COF materials, COF-103-Eth-trans and COF-102-Ant exceed the DOE target of 180 v(STP)/v at 35 bar for methane storage [47].

3.3 Carbon dioxide

In addition to fuel gas, the adsorption of CO_2 by COFs has also been intensively investigated. Carbon dioxide is a major greenhouse gas contributing to the global warming. Over years, capturing CO_2 from industrial emission sources has attracted much interest. COF-1, COF-5, COF-6, COF-8, and COF-10 show CO_2 storage capacity of 230, 870, 310, 630, and 1010 mg g^{-1} at 55 bar and 298 K, respectively, and under the same conditions, COF-102 and COF-103 have higher adsorption capability of 1200 and 1190 mg g^{-1}, respectively [35]. Jiang et al. designed a series of 2D COFs for CO_2 capture and found that TFPA-TAPB-COF with three triarylamine units in the hexagonal pore showed greatly enhanced CO_2 adsorption capacities (33 mg g^{-1}, 1 bar, 298 K and 61 mg g^{-1}, 1 bar, 273 K, respectively), which are 2.5-fold those of TFPB-TAPB-COF [48]. They also developed a series of imine-linked COFs with content-tunable, accessible, and reactive ethynyl groups on the walls of COFs (**Figure 4**). These COFs present an ideal platform for pore surface engineering aimed at anchoring desired functional groups with controllable contents [49]. Their results indicated that the anchoring of amino groups led to an overall increase in CO_2 adsorption. Among these COFs, $[EtNH_2]_{50}$-H_2P-CO exhibited the highest CO_2 adsorption capability of 157 mg g^{-1} at 1 bar and 273 K. The introduction of basic moiety in COFs significantly increases the CO_2 adsorption capacity.

Using the same surface engineering strategy, they further use a conventional imine-linked 2D COF ($[HO]_{X\%}$-H_2P-COF) as a scaffold with porphyrin at the vertices and phenol units on the pore wall of the COFs [31]. A quantitative ring opening reaction between the phenol groups and succinic anhydride decorates the 1D walls with carboxylic acid groups. A series of $[HO_2C]_{X\%}$-H_2P-COFs with controlled carboxylic acid density was obtained. Functionalization of the channel walls with carboxylic acid groups triggers microporosity in $[HO_2C]_{X\%}$-H_2P-COFs, and the CO_2 adsorption capability was increased due to strong dipolar interaction of carboxylic acid groups with CO_2 [31]. Besides, an azine-linked COF, ACOF-1, with high levels of nitrogen and high surface area (1176 m^2 g^{-1}), can store up to 177 mg g^{-1} of CO_2 at 1 bar and 273 K [50].

3.4 Iodine vapor

Nuclear power has high energy density and is gradually becoming into one of the pillars of energy sources in the world. However, disposal of the inevitable nuclear waste generated from nuclear fission still remains great challenges. In this regard, ^{129}I and ^{131}I are one of the most dangerous wastes because of its long radioactive half-life

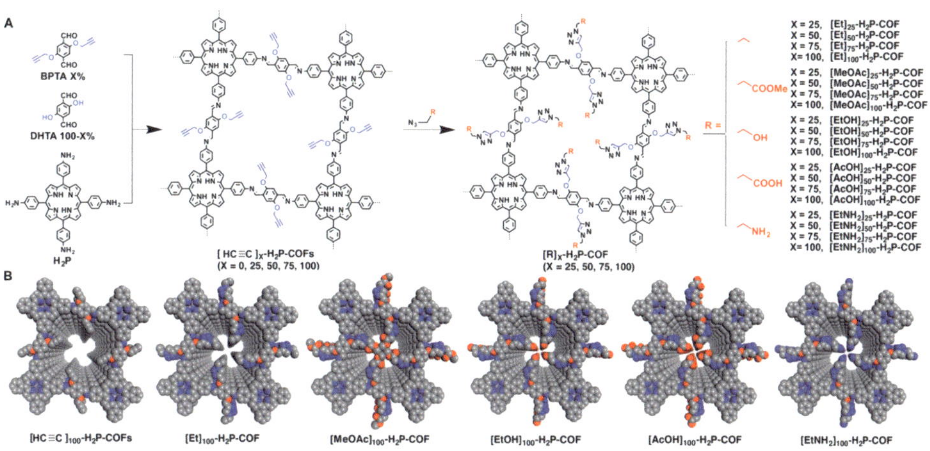

Figure 4.
(a) Schematic of pore surface engineering of imine-linked COFs with various functional groups via click reactions and (b) pore structures of COFs with different functional groups (reproduced from ref. [49], with permission).

(1.57×10^7 years), high volatility, and adverse effects on human being and the environment [51]. Therefore, a complete removal of iodine from wastes is essential for eradicating the long-term pollution. In recent years, the capture of radioactive iodine by using porous materials has attracted much attention. Design of porous materials with high iodine capture capacity holds the key to the innovative technology. Various porous materials, like porous organic cages [52], MOFs [53, 54], activated carbon [55], and porous organic frameworks (POFs) [56, 57] have been studied for iodine adsorption. Although some of them have high surface areas, most of them still exhibit low capacity because of their low accessibility of the pores to iodine molecules.

As a new class of emerging porous materials, the capture of iodine with COFs has been recently explored. Zhao's group reported the construction of a new heteropore COF consisting of two different kinds of micropores with unprecedented shapes [58]. The hollow microspheres were observed for the COF, which led to an extremely high volatile iodine uptake (up to 481 wt%), because iodine can be encapsulated within both the inner cavities of hollow microspheres and the porous shells of the microspheres. In addition to 2D COFs, the capture of iodine by 3D COF has been studied by our group. We designed and synthesized a 3D COF, COF-DL229, which consists of a diamond topology knotted by adamantane units [11]. An eightfold interwoven diamond net with 1D channel structure has been obtained, and the extended π-conjugated linkers are laid on the plane of the rings to form the channel walls. This unique arrangement allows a full exposure of all diphenylamine units to the channels. Our experiments showed that COF-DL229 achieved an uptake capacity of 82.4 wt%, corresponding to an adsorption capability of 4.7 g/g in equilibrium at 75°C. Such high iodine capture capability can be attributed to 1D open channel structure, π-conjugated pore walls that form charge transfer complex with iodine, as well as high surface area of COF-DL229 [11].

3.5 Ammonia

Liquid ammonia is difficult to handle due to its toxicity and corrosiveness. Searching for an adsorbent for efficient storage of ammonia is therefore of significance to solve this problem. A boron-containing COF, COF-10, has exhibited exceptionally high ammonia uptake (15 mol kg^{-1}, 298 K, 1 bar) in comparison with other porous materials [59]. This is higher than other adsorbents. For instance, 13X zeolite has a storage capability of 9 mol kg^{-1}, and Amberlyst 15 exhibits a storage capability of 11 mol kg^{-1} and MCM-41 shows an adsorption capability of 7.9 mol kg^{-1} [60]. Importantly, the captured ammonia can be easily removed only by a simple heating under vacuum.

4. Catalysis

All COFs show an open network structure, which offers accessible channels or nanopores with uniform sizes ranging from angstroms to nanometers for guest molecules. This unique structure is special attractive for catalysis applications. In addition, the organic skeletons of COFs make it easy to be decorated with functional groups. For example, unique optical [61, 62] and optoelectronic [63] properties have been endowed to COFs upon decoration of corresponding functional groups. In particular, the assembly of COFs can be achieved by various designable building blocks, which results in innumerous structural topologies and diversified porosities. High surface area, structural versatility, open channels, and easy modification of COFs make them ideal materials used as catalysts. Compared with traditional activated carbon and zeolites, the "designable" assembly of building blocks enables the spatial separation of multiple catalytic sites in the framework, endowing COFs with cooperative catalysis character and thus enhanced catalyst reactivity [64]. Using COFs as catalysts is a rapidly developing research field. In the following section, we present the recent advances in catalytic COF materials based on the type of catalytic sites, including single functional active sites, bifunctional active sites, and metal nanoparticles (NPs) embedded in pores of COF catalysts. Catalysis using 3D COFs was also discussed separately. The deliberate or incidental synthesis strategies, the stability, the heterogeneity, and the shape/size selectivity of COF catalysts are mainly presented. It is noted that although COFs have also been applied to photocatalytic and electrocatalytic reactions, we do not discuss in this chapter.

4.1 COFs with single functional group

Due to the fact that COFs are composed of organic components, COFs can be easily decorated with molecular catalysts that acquire activities and/or selectivities comparable to their homogeneous analogues. In general, two strategies have been mainly used to construct catalytically active COFs [65]. One strategy involves the postsynthetic integration of catalytic sites into a COF skeleton. With this method, crystalline COFs are firstly synthesized and catalytic sites are then linked within the pores of COFs through chemical reactions. This postsynthetic modification approach can reduce the damage of COF crystallinity due to bulky catalytic sites embedded. The undesired effect of harsh solvothermal conditions on the catalytic sites can also be avoided. So far, this is the most common way to modify COFs. Another one is the bottom-up strategy that uses building blocks bearing catalytic sites to directly construct COFs. However, if a bulky catalytic site is chosen to attach to building blocks, it will be difficult to obtain highly crystalline COFs.

A pioneer work using COFs as catalysts was reported by Wang and co-workers in 2011 [66]. They developed a 2D imine-linked COF, COF-LZU1, and the eclipsed layered-sheet arrangement of the COF gives the distance of eclipsed nitrogen atoms in adjacent layers as ~3.7 Å, which provides an ideal distance for incorporating a variety of metal ions (**Figure 5**).

Pd(II) was embedded into COF-LZU1 through the coordination with nitrogen atoms. The resultant Pd/COF-LZU1 exhibited high catalytic activity for

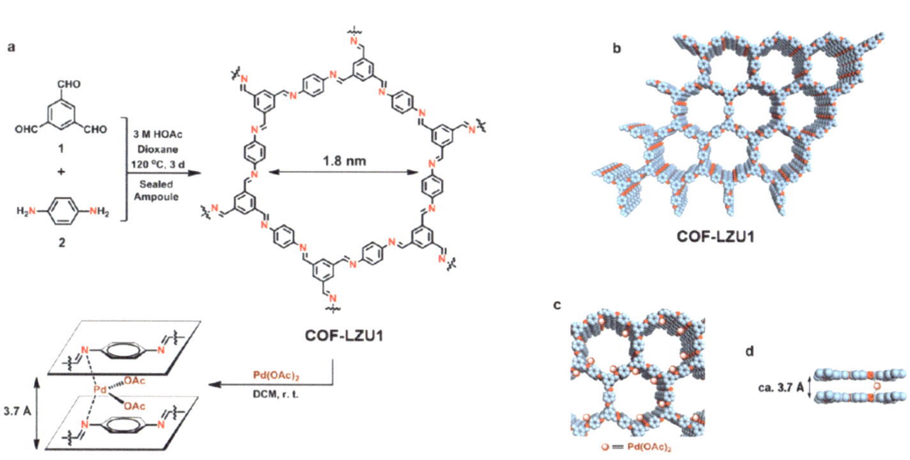

Figure 5.
(a) Schematic of the synthesis of COF-LZU1 and Pd/COF-LZU1. Proposed structures of (b) COF-LZU1 and (c, d) Pd/COF-LZU1 possessing regular microporous channels, simulated with a 2D eclipsed layered-sheet arrangement (reproduced from ref. [66], with permission).

Suzuki-Miyaura coupling reaction. The authors believe that the unique structure of Pd/COF-LZU1 provides efficient access to the catalytic sites and fast mass-transport of the reactants/products, which led to its superior activity in catalyzing the Suzuki-Miyaura reaction. Furthermore, high stability and easy recyclability have also been observed for the catalyst. In a similar fashion, Chen et al. successfully incorporated copper ions (Cu^{2+}) into 2D TAPT-DHTA-COF, in which the hydroxyl groups on the building block can coordinate with Cu^{2+} in combination with imine linkers [67]. The catalytic performances of the resultant Cu^{2+}-containing COFs, termed Cu-COFHX and Cu-COFDMF, have been studied and both COFs exhibited high reactive activity toward selective oxidation of styrene to benzaldehyde. In a similar way, Pd^{2+} ions have been successfully loaded within the layers of COFs through the coordination with either bipyridyl or porphyrin group [68, 69]. The excellent catalytic activity has been observed for the Heck reaction and Suzuki coupling reaction, respectively. In addition, a simple two-step bottom-up synthetic strategy was used to synthesize the molybdenum-doped COF. The as-synthesized Mo-COF was found to be highly active in catalyzing a series of selective oxidation reactions [70]. Based on the principle of reticular chemistry, a new COF, CPF-2, was synthesized through the condensation of 2D four-branched tetraphenylamine porphyrin (TAPP) and the 2D trigonal 1,3,5-triformylbenzene (TFB) [71]. Although X-ray diffraction (XRD) results show that CPF-2 is a crystalline material, the crystal structure is not successfully resolved. Even so, Mn^{2+} and Co^{2+} were successfully coordinated within the porphyrinic framework of CPF-2. The Co(II)-doped material Co-CPF-2 shows high efficiency in heterogeneous catalysis of the aerobic epoxidation of olefins under mild conditions, in which a high conversion (>99%) and epoxide selectivity (93%) have been achieved for the aerobic epoxidation of styrene.

In addition to metal ions, organocatalytic moieties have been introduced into COFs. Jiang's group has performed a series of work to attach organocatalytic sites into COFs. A pore surface engineering strategy was used to modify the properties of COFs. This strategy realizes the control over the number and composition of catalytic sites on the channel walls of COFs, thus providing a basic design principle for the preparation of catalytic COFs. In 2014, they utilized a mesoporous imine-linked porphyrin COF as a scaffold, in which the porphyrin units are located at the vertices and the phenyl groups occupy the edges of tetragonal polygon frameworks [72]. A series of $[HC \equiv C]x$-H_2P-COFs with different alkynyl densities was obtained. Catalytic active pyrrolidine units were then integrated into the pore walls of the $[HC \equiv C]_X$-H_2P-COFs through click reaction. The catalytic activities of $[Pyr]_X$-H_2P-COFs in a Michael addition reaction in aqueous solutions were investigated. It is shown that the COF catalyst features enhanced activity, broad applicability, good recyclability, and high capability under continuous-flow conditions.

They subsequently reported stable COFs, $[HC \equiv C]x$-TPB-DMTP-COFs, with adjustable acetylene density [73]. Anchoring chiral centers and organocatalytic sites onto the channel walls of $[HC \equiv C]x$-TPB-DMTP-COFs was then carried out through click reaction, and chiral organocatalytic $[(S)$-Py$]x$-TPB-DMTP-COFs were obtained. In these COFs, each hexagonal macrocycle can have one, two, or three (*S*)-Py units on average. The organocatalytic sites were integrated into the walls to offer catalytic nanospaces of different sizes (**Figure 6a**). It is found that Michael reactions can be conducted in neat water under mild conditions (25°C and 1 bar), and a 100% conversion, with high isolation yield and high enantioselectivity and diastereoselectivity, has been observed for the addition reaction of cyclohexanone and β-nitrostyrene. This outstanding catalytic performance has also been expanded to different β-nitrostyrenes, illustrating the generality of the catalyst (**Figure 6b**).

Recently, Cui and co-workers have carried out a series of work that used COFs as catalysts. A multivariate strategy was first used to construct chiral COFs. The chiral

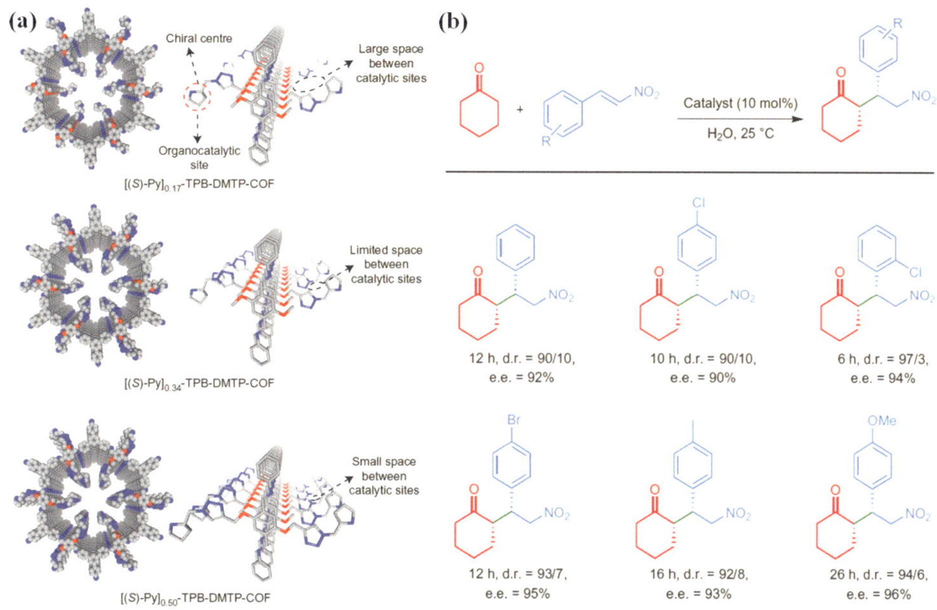

Figure 6.
(a) Channel-wall structure of the chiral organocatalytic COFs. The open structures of the chiral centers and the catalytic sites on the channel walls are shown for different COFs. (b) Scope of reactants. Different β-nitrostyrene derivatives investigated for the Michael reactions catalyzed with chiral COFs, their products, e.e. yields and d.r. values. R, substituent H, Cl, Br, me, or OMe (reproduced from ref. [73], with permission).

COFs exhibit high crystallinity and good stability and thus are used as catalysts for asymmetric catalysis [74]. These 2D chiral COFs showed high catalytic activity in catalyzing asymmetric aminooxylation reaction, aldol reaction, and Diels-Alder reaction. The stereoselectivity and diastereoselectivity for the reactions are found to be comparable to or surpassing their corresponding homogeneous counterparts. A metal-directed synthesis strategy has also been developed by Cui's group, in which chiral Zn(salen)-based COFs have been obtained by the imine-condensations of enantiopure 1,2-diaminocyclohexane and C_3-symmetric trisalicylaldehydes with one or no 3-tert-butyl group (**Figure 7a**) [75]. Metal ions can be loaded within the Zn(salen) modules of the COFs by using an ion exchange process. The loading of metals did not change high crystallinity and porosity of the COF. The catalytic performance of the resultant COFs was detected by a series of reactions, including asymmetric cyanation of aldehydes, Diels-Alder reaction, alkene epoxidation, and epoxide ring-opening reaction. The excellent catalytic activity and selectivity have been observed for these reactions (**Figure 7b**).

Very recently, they also demonstrated that chiral COFs can be prepared from achiral organic precursors by chiral catalytic induction. Nine 2D chiral COFs were synthesized under solvothermal conditions by the imine condensations of 1,3,5-triformylphloroglucinol (Tp) and diamine or triamine linkers in the presence of catalytic amount of (*R*)- or (*S*)-1-phenylethylamine [76]. Using CCOF-TpTab as a heterogeneous catalyst, high reactive activity has been observed in catalyzing asymmetric Henry reaction. In this respect, our group has successfully introduced ionic liquids into the COF framework by using a postsynthetic strategy [77]. The immobilization of ionic liquids onto Py-COFs was directly accomplished through the Williamson ether reaction between the phenol group in $[HO]_{X\%}$-Py-COFs and (2-bromoethyl)triethylammonium bromide ionic liquid. Among these COFs, $[Et_4NBr]_{50\%}$-Py-COF showed a strong CO_2 adsorption capability of 164.6 mg g^{-1} at 273 K and 1 bar. The catalytic transformation of CO_2 into formamides was detected

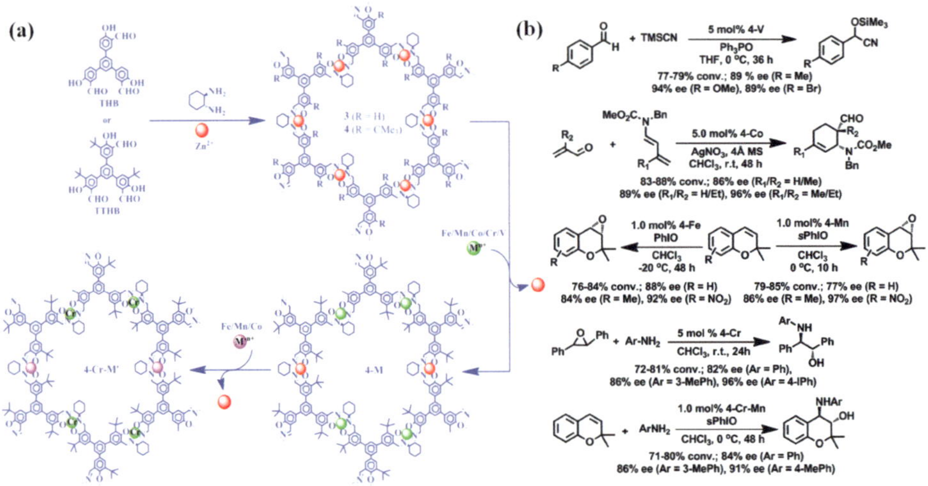

Figure 7.
(a) Synthesis of the chiral COFs. (b) Asymmetric reactions catalyzed by the CCOFs (reproduced from ref. [75], with permission).

and the $[Et_4NBr]_{50\%}$-Py-COF showed high catalytic activity even if the reaction was carried out at ambient temperature and pressure. We believe that the enrichment of CO_2 into the COF nanopores may lead to its outstanding catalytic ability.

Using a bottom-up strategy to construct functional COFs is relatively difficult because it must simultaneously satisfy requirement of crystallinity and functionality. So far, only several attempts have shown to be successful for the synthesis of catalytic COFs. A facile strategy for the direct construction of chiral-functionalized COFs using chiral (*S*)-4,4′-(2-(pyrrolidin-2-yl)-1H-benzo[d]imidazole-4,7-diyl) dianiline (**Figure 8a**) as a building block has been achieved by Wang and co-workers [78]. Two chiral COFs, LZU-72 and LZU-76, were prepared based on this chiral building block (**Figure 8b**). Their results showed that these two COFs are structurally robust and highly active as heterogeneous organocatalysts. LZU-76 can catalyze asymmetric aldol reaction to produce the corresponding products with excellent enantioselectivity (88.4:11.6–94.0:6.0 e.r.) (**Figure 8c**).

A sulfonated building block, 2,5-diaminobenzenesulfonic acid, was also used to construct an acidic COF, termed TFP-DABA, together with 1,3,5-triformylphloroglucinol [79]. The as-synthesized TFP-DABA was found to be a highly efficient catalyst for fructose conversion. The remarkable yields (97% for 5-hydroxymethylfurfural (HMF) and 65% for 2,5-diformylfuran), good chemoselectivity and recyclability have been observed for this catalyst. Cui and co-workers [80] reported the synthesis of two chiral COFs (CCOF-1 and CCOF-2) by direct condensations of enantiopure TADDOL-derived tetraaldehydes with 4,4′-diaminodiphenylmethane (4,4′-DADPM). After optimization of the reaction conditions, CCOF-1 in the presence of excess $Ti(OiPr)_4$ was found to be highly active for the addition reaction of Et_2Zn to aromatic aldehydes to produce secondary alcohols.

In some cases, COFs themselves can serve as efficient catalysts for some reactions. CTFs are a special class of COFs constructed through the reversible ionothermal trimerization of aromatic nitriles. The inherent basic feature of triazine materials means that they can be used as basic catalysts. The efficient catalytic activities of the high-surface-area triazine frameworks have been clearly demonstrated in the cycloaddition of CO_2 to epichlorohydrin [81]. Besides, Jiang and co-workers [82] found that a π-electronic COF can serve as heterogeneous catalyst. The use of 1D π-walls as catalytic beds can facilitate the reaction of Diels-Alder

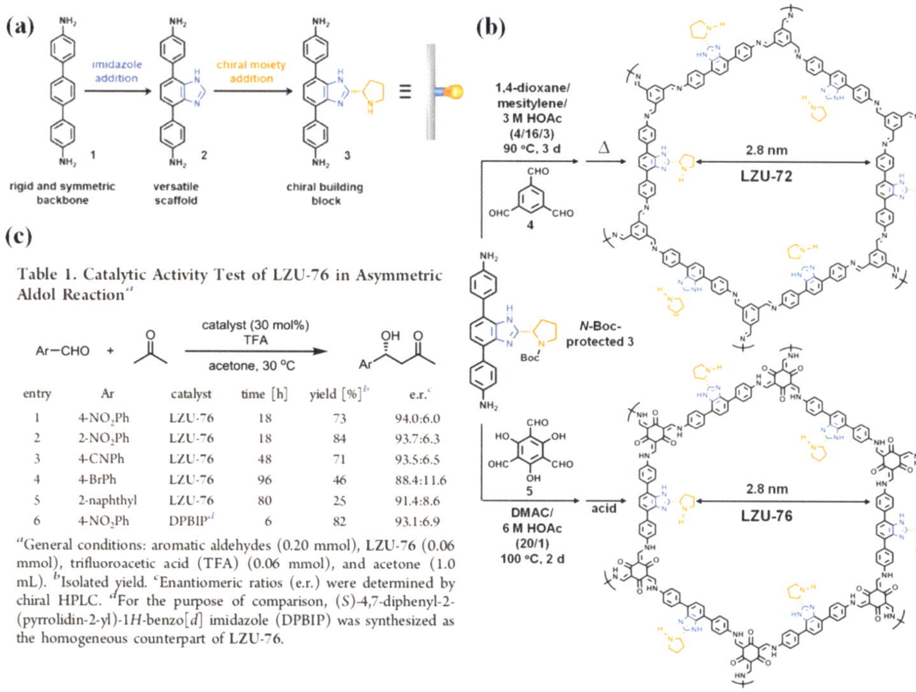

Table 1. Catalytic Activity Test of LZU-76 in Asymmetric Aldol Reaction[a]

entry	Ar	catalyst	time [h]	yield [%][b]	e.r.[c]
1	4-NO_2Ph	LZU-76	18	73	94.0:6.0
2	2-NO_2Ph	LZU-76	18	84	93.7:6.3
3	4-CNPh	LZU-76	48	71	93.5:6.5
4	4-BrPh	LZU-76	96	46	88.4:11.6
5	2-naphthyl	LZU-76	80	25	91.4:8.6
6	4-NO_2Ph	DPBIP[d]	6	82	93.1:6.9

[a]General conditions: aromatic aldehydes (0.20 mmol), LZU-76 (0.06 mmol), trifluoroacetic acid (TFA) (0.06 mmol), and acetone (1.0 mL). [b]Isolated yield. [c]Enantiomeric ratios (e.r.) were determined by chiral HPLC. [d]For the purpose of comparison, (*S*)-4,7-diphenyl-2-(pyrrolidin-2-yl)-1*H*-benzo[*d*] imidazole (DPBIP) was synthesized as the homogeneous counterpart of LZU-76.

Figure 8.
(a) Design of chiral building block. (b) Synthesis of chiral LZU-72 and LZU-76. (c) Catalytic activity test of LZU-76 in asymmetric aldol reaction (reproduced from ref. [78], with permission).

reactions. Their result indicated that the reaction can proceed effectively at room temperature and at 1 bar.

4.2 COFs with bifunctional reactive sites

It is of great interest to combine two different active sites in one substrate as in many cases, the reaction can be facilitated remarkably, especially for cascade/tandem/one-pot reactions. However, there still remains a challenge to design bifunctional catalysts. Up to now, only few such systems of this sort have been reported.

The first successful incorporation of bifunctional catalytic sites in the crystalline porous COF (2,3-DhaTph) was developed by Banerjee et al. [83]. They report the synthesis of a bifunctional catechol-porphyrin COF, containing both acidic and basic sites, which can act as a heterogeneous catalyst for one-pot acid-catalyzed deprotection of acetal groups followed by base-catalyzed Knoevenagel condensation reaction. The reaction of benzaldehydedimethylacetal with malononitrile in a mixture of toluene and water at 80°C gives the desired 2-benzylidenemalononitrile product with an excellent isolated yield of 96% in the presence of 2,3-DhaTph. Our group used 2,2-bipyridyl-5,5-dialdehyde (2,2'-BPyDCA) as building blocks to construct a series of 2D imine-based 2D COFs (X% BPy COF) [84]. It is known that imine at the layers of 2D COFs can provide nitrogen atoms to coordinate with metal ions [66]. This means that X% BPy COFs contain two different types of ligands, namely imine and bipyridine in one COF. Through a programmed synthetic procedure, big-sized Rh(I) was first coordinated with bipyridine on the walls of the COFs and Pd(II) was then coordinated with the nitrogen atoms of the imine within the layers of X% BPy COFs. The resultant bimetallic Rh^{I}/Pd^{II}@X% BPy COFs were used as catalysts for a cascade reaction. The reaction of phenylboronic acid with benzaldehyde gives a final product of benzophenone. Our investigation has revealed that Rh^{I} catalyzed the addition reaction of phenylboronic acid with benzaldehyde to product diphenylmethanol, and the intermediate was then oxidized into benzophenone catalyzed by Pd^{II}. A total 90% isolation yield was obtained, showing an excellent catalytic activity of the bimetallically docked 2D COFs in the one-pot addition-oxidation cascade reaction. The high surface area, controllable metal-loading content, and predesigned active sites make COFs ideal candidates for using as heterogeneous catalysts in a wide range of reactions. In addition, Ma et al. [85] has developed a new strategy to decorate functional active sites into COFs. Surface Lewis acid sites are anchored on the walls of COF-TpBpy through the metalation of Cu^{2+} with building blocks, namely bipyridine units, and linear polymers with excellent flexibility are threaded throughout the channels of the COF via *in situ* radical polymerization of ionic monomers (**Figure 9**). Due to the flexibility of linear polymers, it is possible for polymers in close proximity surface Lewis acid sites, which make them easy to cooperate with the co-catalyst, Br^{-}, of the polymers. This special structure can catalyze the cycloaddition of the epoxides and CO_2 efficiently.

4.3 COFs with metal nanoparticles (NPs) entrapped

The inherent host-guest chemistry enables the implementation of desired substances by filling the pores with various guest molecules. Metal NPs have demonstrated excellent catalytic activity in many chemical reactions. Therefore, the doping of COFs with metal NPs is of significant interest for application in heterogeneous catalysis. We can predict the advantages of metal NP-loaded COFs: (1) high surface area of COFs allows for more active sites being embedded, which can facilitate chemical reactions; (2) postsynthetic modification or prior introduction of ligands makes it easy to anchor metal NPs; (3) the pore channels in COFs are well isolated, and thus, the aggregation of entrapped NPs can be hampered and metal NP growth can be confined to an certain extent.

The first example to load metal NPs was carried out by Banerjee and co-workers in 2014 [86]. A highly stable, porous, and crystalline COF (TpPa-1) was first prepared and the $Au(_0)$@TpPa-1 catalyst was synthesized via the solution infiltration method (**Figure 10**). The as-synthesized $Au(_0)$@TpPa-1 catalyst is stable and shows superior reactivity for nitrophenol reduction reaction. With the same strategy, they further reported the incorporation of $Pd(_0)$ NPs into the pores of TpPa-1 COF [87]. The as-synthesized $Pd(_0)$@TpPa-1 has proven highly active toward the Cu-free Sonogashira, Heck, and sequential one-pot Heck-Sonogashira cross-coupling reactions. The COF is also highly stable under the reaction conditions because of

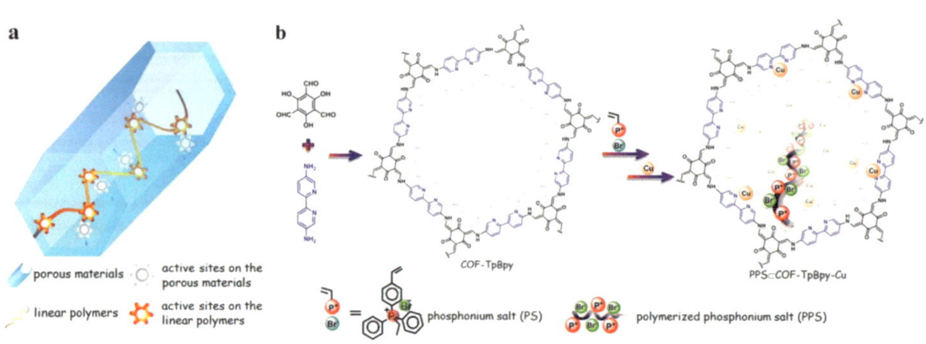

Figure 9.
(a) The concept of heterogeneous concerted catalysis between active sites on the porous materials and highly flexible linear polymers and (b) schematic of PPS@COF-TpBpy-Cu synthesis and structures of COF-TpBpy and PPS@COF-TpBpy-Cu (reproduced from ref. [91], with permission).

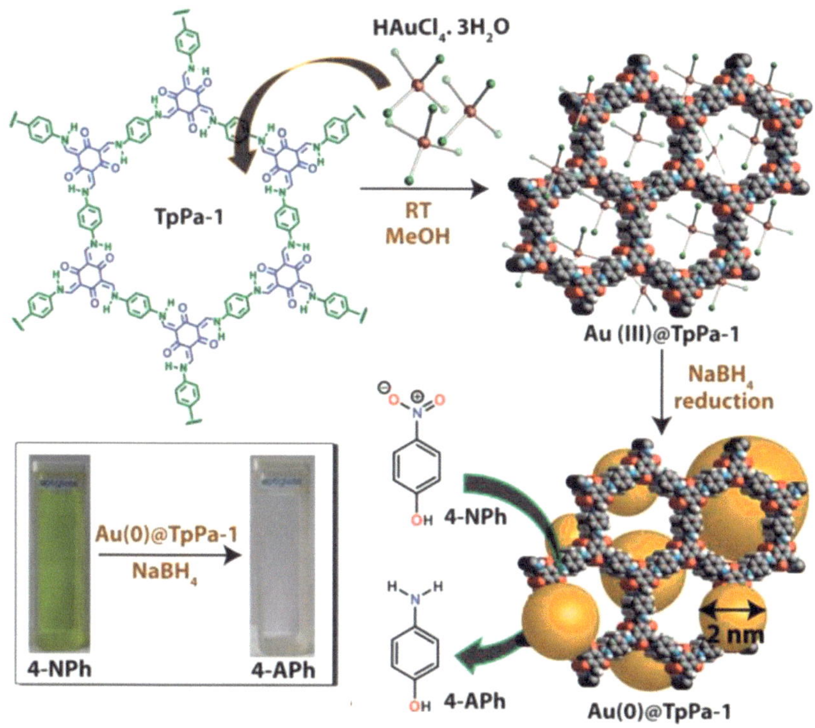

Figure 10.
Synthesis of the Au($_0$)@TpPa-1 catalyst using the solution infiltration method for nitrophenol reduction reaction. Inset image: The optical images of the color change observed for the conversion of 4-nitrophenol to 4-aminophenol after the addition of Au($_0$)@TpPa-1 (reproduced from ref. [86], with permission).

its negligible metal leaching, nonsintering behavior, and good recyclability. In 2017, they further developed a novel synthetic strategy by choosing a predesigned metal-anchored building block for in situ generation of metal (Pd) NPs in the stable, porous, and crystalline COF without using conventional reducing agents [88]. It was revealed that the breakage of the Pd-N bond of the Bpy-$PdCl_2$ complex results in the formation of Pd($_0$) NPs. This hybrid COF is found to be an excellent reusable heterogeneous catalyst for the synthesis of 2-substituted benzofurans from 2-bromophenols and terminal alkynes via a tandem process.

It is of great interest to load metal NPs with desired shape and size into a COF matrix. Such composite has been produced by an encapsulation strategy. The polyvinylpyrrolidone (PVP)-modified Au NPs with desired shape and size have been successfully loaded within the nanopores of TAPB-DMTP COF during its synthesis process [89]. It is revealed that the incorporation of Au NPs does not obviously change the crystallinity and pore structures of the COF matrices. The Au NP-loaded TAPB-DMTP COF displays recyclable catalytic performance for reduction of 4-nitrophenol.

A facile single-step loading of Pd($_0$) NPs into an amphiphilic triazine COF has been carried out by Vaidhyanathan et al. [90]. The catalytic performance was evaluated by Heck coupling and C-C coupling reactions. The COF catalyst exhibits high catalytic activity and good recyclability for Heck couplings with no sign of catalyst leaching. As an oxidation catalyst, CO was 100% conversed to CO_2 at 150°C with no loss of activity with time. Besides, COFs have been found to be ideal templates for the synthesis of ultrafine metal NPs [91]. Both size and dispersity of NPs into COF's matrix could be well controlled. In this regard, a thioether-containing building block has been designed to construct task-specific COF (Thio-COF). The

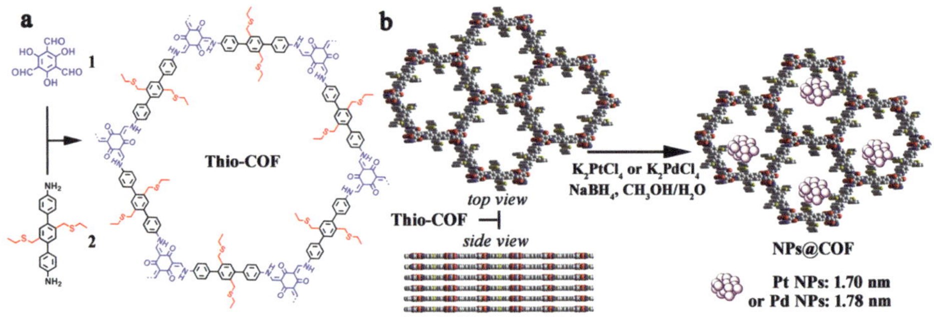

Figure 11.
(a) Synthesis of Thio-COF and (b) schematic representation of the synthesis of Thio-COF supported PtNPs@COF and PdNPs@COF. Top and side views of the energy-minimized models of Thio-COF (yellow, S; blue, N; gray, C; red, O) are shown in (b) (reproduced from ref. [85], with permission).

appearance of thioether confined the growth of metal NPS and the ultrafine NPs (1.70 nm for Pt and 1.78 nm for Pd NPs) were obtained inside the porosity of the COF (**Figure 11**). The excellent catalytic activities have been observed in nitrophenol reduction and Suzuki-Miyaura coupling reaction under mild conditions.

4.4 Catalysis with 3D COFs

In 2014, Yan and co-workers used 3D COFs as catalyst for chemical reactions [92]. Two 3D COFs (BF-COF-1 and BF-COF-2) have been synthesized based on the condensation reaction of 1,3,5,7-tetraaminoadamantane and 1,3,5-triformylbenzene or triformylphloroglucinol. Both BF-COF-1 and BF-COF-2 exhibited high catalytic activity toward Knoevenagel condensation reactions (96% for BF-COF-1 and 98% for BF-COF-2, respectively). Interestingly, the COFs exhibit highly efficient size selectivity due to the pore size of the COFs. For small-sized reactants, like benzaldehyde and malononitrile, the 3D COFs exhibited very good conversion up to 99%, whereas for the large reactants, the reaction proceeded much sluggishly with very poor yield. Constructing COFs by using two different types of reversible covalent bonds has been achieved by Qiu's group [93]. The two linkages, boroxine and imine, provide acidic and basic sites, separately, in one 3D COF material. The double-linked COF can serve as excellent acid-base catalysts for one-pot cascade reactions. The high conversion of 100% for benzaldehyde dimethyl acetal is achieved, and the formed intermediate, benzaldehyde, is reacted through consecutive Knoevenagel condensation with malononitrile to give the final product, benzylidene malononitrile in high yields (98% for DLCOF-1 and 96% for DL-COF-2). Wang et al. described a dynamic 3D COF LZU-301 used as a Lewis-base catalyst for the Knoevenagel condensation between malononitrile and three aromatic aldehydes. An obvious size selectivity effect has been observed for different sized aldehydes [94].

5. Outlook and conclusions

Covalent organic frameworks (COFs) represent an emerging class of crystalline porous materials. Owing to their unique structural features and recent interesting exploitation of their properties, these materials undoubtedly show significant potential and advantages in gas storage and separation, catalysis, optoelectricity, and so on. Over the past 10 years, great progress has been made; however, the investigation of COF materials is still in its infancy, and COF chemistry still faces great challenge in practical applications.

A big challenge in the COF synthesis is the exploration of new synthesis strategy. The successful synthesis of crystalline COFs has long been an intractable problem. So far, there is still lack of universal regulation to direct the construction of crystalline COFs. Therefore, there is a clear need to expand the synthetic possibilities in different ways. In-depth understanding of the formation mechanism would provide insightful information for the COF synthesis. In this respect, ionic liquids have been attempted to act as reaction media and exhibited obvious advantage over traditional organic solvents. Given the fact that there are at least a million binary ionic liquids and potentially more than 10^{18} ternary ionic liquids (compared with only about 600 molecular solvents) [17], the use of ionic liquids would open up many new possibilities in the preparation of COF materials with highly periodic and robust frameworks. Furthermore, synthesis of COFs is now limited to the laboratory, and it is also a challenge to synthesize COFs on an industrial scale for prospective practical applications. Also, how to find a new synthetic route to reduce the cost of COFs is also a difficult task and far in the future.

Another challenge in the COF field is the resolution of their crystalline structure. So far, the structure of COFs is expected to be analyzed based on the given PXRD patterns combined with computational simulation technique. However, in many cases, due to the partial crystallinity of COFs, the overall structural regularity is poor. Apart from searching for an effective synthesis route to construct highly crystalline COFs, efforts for the installation of crystalline structural database of COFs might be encouraged. Besides, how to further improve the chemical/thermal stability of COFs so that these materials can be used under very harsh conditions (e.g., strong acidity or alkalinity, high temperature, and high overpotential) is another important aspect that requires rapid development.

So far, most of the reported COFs are 2D layered crystalline structures, i.e., 2D COFs. In comparison with the effects governed by DCC, the π-π stacking interaction between layers of 2D COFs is relatively weak to construct highly periodic and porous COF materials. Therefore, it is also of significance to develop more 3D COFs. Take catalytic COFs as an example, it is of interest to design COFs with versatile inner cavities, which can provide active sites on the pore surface and transport reactants/products to or from inner reactive vessels.

Acknowledgements

The authors gratefully acknowledge financial support from the National Natural Science Foundation of China (21473196), the start-up grant of Hainan University (KYQD(ZR)-1856), and the collaborative innovation fund of Tianjin University-Hainan University joint project (HDTDU201801).

Conflict of interest

The authors declare no competing financial interest.

Author details

Lifeng Deng, Junfeng Zhang and Yanan Gao*
Key Laboratory of Ministry of Education for Advanced Materials in Tropical Island Resources, Hainan University, Haikou, China

*Address all correspondence to: ygao@hainu.edu.cn

References

[1] Long JR, Yaghi OM. The pervasive chemistry of metal-organic frameworks. Chemical Society Reviews. 2009;**38**:1213-1214. DOI: 10.1039/b903811f

[2] Côté AP, Benin AI, Ockwig NW, O'Keeffe M, Matzger AJ, Yaghi OM. Porous, crystalline, covalent organic frameworks. Science. 2005;**310**: 1166-1170. DOI: 10.1126/science.1120411

[3] Feng X, Ding XS, Jiang DL. Covalent organic frameworks. Chemical Society Reviews. 2012;**41**:6010-6022. DOI: 10.1039/c2cs35157a

[4] Ding SY, Wang W. Covalent organic frameworks (COFs): From design to applications. Chemical Society Reviews. 2013;**42**:548-568. DOI: 10.1039/c2cs35072f

[5] Rowan SJ, Cantrill SJ, Cousins GRL, Sanders JKM, Stoddart JF. Dynamic covalent chemistry. Angewandte Chemie, International Edition. 2002;**41**:898-952. DOI: 10.1002/1521-3773(20020315)41:6<898::AID-ANIE898>3.0.CO;2-E

[6] Lin GQ, Ding HM, Chen RF, Peng ZK, Wang BS, Wang C. 3D porphyrin-based covalent organic frameworks. Journal of the American Chemical Society. 2017;**139**:8705-8709. DOI: 10.1021/jacs.7b04141

[7] Ding XS, Guo J, Feng X, Honsho Y, Guo JD, Seki S, et al. Synthesis of metallophthalocyanine covalent organic frameworks that exhibit high carrier mobility and photoconductivity. Angewandte Chemie, International Edition. 2011;**50**:1289-1293. DOI: 10.1002/anie.201005919

[8] Tilford RW, Gemmill WR, zur Loye HC, Lavigne JJ. Facile synthesis of a highly crystalline, covalently linked porous boronate network. Chemistry of Materials. 2006;**18**:5296-5301. DOI: 10.1021/cm061177g

[9] Li LH, Feng XL, Cui XH, Ma YX, Ding SY, Wang W. Salen-based covalent organic framework. Journal of the American Chemical Society. 2017;**139**:6042-6045. DOI: 10.1021/jacs.7b01523

[10] Fang QR, Wang JH, Gu S, Kaspar RB, Zhuang ZB, Zheng J, et al. 3D porous crystalline polyimide covalent organic frameworks for drug delivery. Journal of the American Chemical Society. 2015;**137**:8352-8355. DOI: 10.1021/jacs.5b04147

[11] Wang C, Wang Y, Ge RL, Song XD, Xing XQ, Jiang QK, et al. A 3D covalent organic framework with exceptionally high iodine capture capability. Chemistry--A European Journal. 2018;**24**:585-589. DOI: 10.1002/chem.201705405

[12] Feng X, Chen L, Dong YP, Jiang DL. Porphyrin-based two-dimensional covalent organic frameworks: Synchronized synthetic control of macroscopic structures and pore parameters. Chemical Communications. 2011;**47**:1979-1981. DOI: 10.1039/c0cc04386a

[13] Kuhn P, Antonietti M, Thomas A. Porous, covalent triazine-based frameworks prepared by ionothermal synthesis. Angewandte Chemie, International Edition. 2008;**47**: 3450-3453. DOI: 10.1002anie.200705710

[14] Bojdys MJ, Jeromenok J, Thomas A, Antonietti M. Rational extension of the family of layered, covalent, triazine-based frameworks with regular porosity. Advanced Materials. 2010;**22**:2202-2205. DOI: 10.1002/adma.200903436

[15] Guan XY, Ma YC, Li H, Yusran Y, Xue M, Fang QR, et al. Fast, ambient temperature and pressure ionothermal synthesis of three-dimensional covalent organic frameworks. Journal of the American Chemical Society. 2018;**140**:4494-4498. DOI: 10.1021/jacs.8b01320

[16] Hallett JP, Welton T. Room-temperature ionic liquids: Solvents for synthesis and catalysis. 2. Chemical Reviews. 2011;**111**:3508-3576. DOI: 10.1021/cr1003248

[17] Cooper ER, Andrews CD, Wheatley PS, Webb PB, Wormald P, Morris RE. Nature. Ionic liquids and eutectic mixtures as solvent and template in synthesis of zeolite analogues. 2004;**430**:1012-1016. DOI: 10.1038/nature02860

[18] Lin ZJ, Slawin AMZ, Morris RE. Chiral induction in the ionothermal synthesis of a 3-D coordination polymer. Journal of the American Chemical Society. 2007;**129**:4880-4881. DOI: 10.1021/ja070671y

[19] Antonietti M, Kuang D, Smarsly B, Zhou Y. Ionic liquids for the convenient synthesis of functional nanoparticles and other inorganic nanostructures. Angewandte Chemie, International Edition. 2004;**43**:4988-4992. DOI: 10.1002/anie.200460091

[20] Ni Z, Masel RI. Rapid production of metal-organic frameworks via microwave-assisted solvothermal synthesis. Journal of the American Chemical Society. 2006;**128**: 12394-12395. DOI: 10.1021/ja0635231

[21] Campbell NL, Clowes R, Ritchie LK, Cooper AI. Rapid microwave synthesis and purification of porous covalent organic frameworks. Chemistry of Materials. 2009;**21**:204-206. DOI: 10.1021/cm802981m

[22] Ritchie LK, Trewin A, Reguera-Galan A, Hasell T, Cooper AI. Synthesis of COF-5 using microwave irradiation and conventional solvothermal routes. Microporous and Mesoporous Materials. 2010;**132**:132-136. DOI: 10.1016/j.micromeso.2010.02.010

[23] El-Kaderi HM, Hunt JR, Mendoza-Cortes JL, Cote AP, Taylor RE, O'Keeffe M, et al. Designed synthesis of 3D covalent organic frameworks. Science. 2007;**316**:268-272. DOI: 10.1126/science.1139915

[24] Smith BJ, Dichtel WR. Mechanistic studies of two-dimensional covalent organic frameworks rapidly polymerized from initially homogenous. Journal of the American Chemical Society. 2014;**136**:8783-8789. DOI: 10.1021/ja5037868

[25] Côté AP, El-Kaderi HM, Furukawa H, Hunt JR, Yaghi OM. Reticular synthesis of microporous and mesoporous 2D covalent organic frameworks. Journal of the American Chemical Society. 2007;**129**: 12914-12915. DOI: 10.1021/ja0751781

[26] Spitler EL, Colson JW, Uribe-Romo FJ, Wol AR, Giovino MR, Saldivar A, et al. Lattice expansion of highly oriented 2D phthalocyanine covalent organic framework films. Angewandte Chemie, International Edition. 2012;**51**:2623-2627. DOI: 10.1002/anie.201107070

[27] Rabbani MG, Sekizkardes AK, Kahveci Z, Reich TE, Ding RS, El-Kaderi HM. A 2D mesoporous imine-linked covalent organic framework for high pressure gas storage applications. Chemistry - A European Journal. 2013;**19**:3324-3328. DOI: 10.1002/chem.201203753

[28] Uribe-Romo FJ, Hunt JR, Furukawa H, Klöck C, O'Keeffe M, Yaghi OM. A crystalline imine-linked 3-D porous covalent organic framework. Journal of the American Chemical Society.

2009;**131**:4570-4571. DOI: 10.1021/ja8096256

[29] Zhang YB, Su J, Furukawa H, Yun YF, Gándara F, Duong A, et al. Single-crystal structure of a covalent organic framework. Journal of the American Chemical Society. 2013;**135**: 16336-16339. DOI: 10.1021/ja409033p

[30] Baldwin LA, Crowe JW, Pyles DA, McGrier PL. Metalation of a mesoporous three-dimensional covalent organic framework. Journal of the American Chemical Society. 2016;**138**:15134-15137. DOI: 10.1021/jacs.6b10316

[31] Huang N, Chen X, Krishna R, Jiang DL. Two-dimensional covalent organic frameworks for carbon dioxide capture through channel-wall functionalization. Angewandte Chemie, International Edition. 2015;**54**:2986-2990. DOI: 10.1002/anie.201411262

[32] Cao DP, Lan JH, Wang WC, Smit B. Lithium-doped 3D covalent organic frameworks: High-capacity hydrogen storage materials. Angewandte Chemie, International Edition. 2009;**48**: 4730-4733. DOI: 10.1002/anie.200900960

[33] Guo JH, Zhang H, Miyamoto Y. New Li-doped fullerene-intercalated phthalocyanine covalent organic frameworks designed for hydrogen storage. Physical Chemistry Chemical Physics. 2013;**15**:8199-8207. DOI: 10.1039/C3CP50492A

[34] Lan JH, Cao DP, Wang WC, Smit B. Doping of alkali, alkaline-earth, and transition metals in covalent-organic frameworks for enhancing CO_2 capture by first-principles calculations and molecular simulations. ACS Nano. 2010;**4**:4225-4237. DOI: 10.1021/nn100962r

[35] Tilford RW, Mugavero SJ III, Pellechia PJ, Lavigne JJ. Tailoring microporosity in covalent organic frameworks. Advanced Materials. 2008;**20**:2741-2746. DOI: 10.1002/adma.200800030

[36] Yu JT, Chen Z, Sun J, Huang ZT, Zheng QY. Cyclotricatechylene based porous crystalline material: Synthesis and applications in gas storage. Journal of Materials Chemistry. 2012;**22**: 5369-5373. DOI: 10.1039/C2JM15159F

[37] Furukawa H, Yaghi OM. Storage of hydrogen, methane, and carbon dioxide in highly porous covalent organic frameworks for clean energy applications. Journal of the American Chemical Society. 2009;**131**:8875-8883. DOI: 10.1021/ja9015765

[38] Ben T, Pei CY, Zhang DL, Xu J, Deng F, Jing XF, et al. Gas storage in porous aromatic frameworks (PAFs). Energy & Environmental Science. 2011;**4**: 3991-3999. DOI: 10.1039/C1EE01222C

[39] Zeng YF, Zou RY, Luo Z, Zhang HC, Yao X, Ma X, et al. Covalent organic frameworks formed with two types of covalent bonds based on orthogonal reactions. Journal of the American Chemical Society. 2015;**137**:1020-1023. DOI: 10.1021/ja510926w

[40] Furukawa H, Yaghi OM. A covalent organic framework that exceeds the DOE 2015 volumetric target for H_2 uptake at 298 K. Journal of Physical Chemistry Letters. 2012;**3**:2671-2675. DOI: 10.1021/jz301000m

[41] Han SS, Furukawa H, Yaghi OM, Goddard WA. Covalent organic frameworks as exceptional hydrogen storage materials. Journal of the American Chemical Society. 2008;**130**:11580-11581. DOI: 10.1021/ja803247y

[42] Klontzas E, Tylianakis E, Froudakis GE. Hydrogen storage in 3D covalent organic frameworks. A multiscale theoretical investigation. Journal of

Physical Chemistry C. 2008;**112**: 9095-9098. DOI: 10.1021/jp711326g

[43] Klontzas E, Tylianakis E, Froudakis GE. Designing 3D COFs with enhanced hydrogen storage capacity. Nano Letters. 2010;**10**:452-454. DOI: 10.1021/nl903068a

[44] Kim D, Jung DH, Kim KH, Guk H, Han SS, Choi K, et al. Pillared covalent organic frameworks with balanced volumetric and gravimetric hydrogen uptake. Journal of Physical Chemistry C. 2012;**116**:1479-1484. DOI: 10.1021/jp2080622

[45] Furukawa H, Ko N, Go YB, Aratani N, Choi SB, Choi E, et al. Ultrahigh porosity in metal-organic frameworks. Science. 2010;**329**:424-428. DOI: 10.1126/science.1192160

[46] Lan JH, Cao DP, Wang WC. High uptakes of methane in Li-doped 3D covalent organic frameworks. Langmuir. 2010;**26**:220-226. DOI: 10.1021/la9020383

[47] Mendoza-Cortes JL, Pascal TA, Goddard WA III. Design of covalent organic frameworks for methane storage. The Journal of Physical Chemistry. A. 2011;**115**:13852-13857. DOI: 10.1021/jp209541e

[48] Zhai LP, Huang N, Xu H, Chen QH, Jiang DL. A backbone design principle for covalent organic frameworks: The impact of weakly interacting units on CO_2 adsorption. Chemical Communications. 2017;**53**:4242-4245. DOI: 10.1039/C7CC01921A

[49] Huang N, Krishna R, Jiang DL. Tailor-made pore surface engineering in covalent organic frameworks: Systematic functionalization for performance screening. Journal of the American Chemical Society. 2015;**137**:7079-7082. DOI: 10.1021/jacs.5b04300

[50] Li ZP, Feng X, Zou YC, Zhang YW, Xia H, Liu XM, et al. A 2D azine-linked covalent organic framework for gas storage applications. Chemical Communications. 2014;**50**:13825-13828. DOI: 10.1039/c4cc05665e

[51] Sava DF, Rodriguez MA, Chapman KW, Chupas PJ, Greathouse JA, Crozier PS, et al. Capture of volatile iodine, a gaseous fission product, by zeolitic imidazolate framework-8. Journal of the American Chemical Society. 2011;**133**:12398-12401. DOI: 10.1021/ja204757x

[52] Hasell T, Schmidtmann M, Cooper AI. Molecular doping of porous organic cages. Journal of the American Chemical Society. 2011;**133**:14920-14923. DOI: 10.1021/ja205969q

[53] Yin Z, Wang QX, Zeng MH. Iodine release and recovery, influence of polyiodide anions on electrical conductivity and nonlinear optical activity in an interdigitated and interpenetrated bipillared-bilayer metal-organic framework. Journal of the American Chemical Society. 2012;**134**:4857-4863. DOI: 10.1021/ja211381e

[54] Abrahams BF, Moylan M, Orchard SD, Robson R. Zinc saccharate: A robust, 3D coordination network with two types of isolated, parallel channels, one hydrophilic and the other hydrophobic. Angewandte Chemie, International Edition. 2003;**42**: 1848-1851. DOI: 10.1002/anie.200250633

[55] Nguyen NV, Jeong J, Shin D, Kim BS, Lee J, Pandey BD. Simultaneous recovery of gold and iodine from the waste rinse water of the semiconductor industry using activated carbon. Materials Transactions. 2012;**53**:760-765. DOI: 10.2320/matertrans.M2012009

[56] Yan ZJ, Yuan Y, Tian YY, Zhang DM, Zhu GS. Highly efficient enrichment of volatile iodine by charged porous aromatic frameworks with three

sorption sites. Angewandte Chemie, International Edition. 2015;**54**:12733-12737. DOI: 10.1002/anie.201503362

[57] Comotti A, Bracco S, Ben T, Qiu SL, Sozzani P. Molecular rotors in porous organic frameworks. Angewandte Chemie, International Edition. 2014;**53**:1043-1047. DOI: 10.1002/anie.201309362

[58] Yin ZJ, Xu SQ, Zhan TG, Qi QY, Wu ZQ, Zhao X. Ultrahigh volatile iodine uptake by hollow microspheres formed from a heteropore covalent organic framework. Chemical Communications. 2017;**53**:7266-7269. DOI: 10.1039/C7CC01045A

[59] Doonan CJ, Tranchemontagne DJ, Glover TG, Hunt JR, Yaghi OM. Exceptional ammonia uptake by a covalent organic framework. Nature Chemistry. 2010;**2**:235-238. DOI: 10.1038/NCHEM.548

[60] Helminen J, Helenius J, Paatero E. Adsorption equilibria of ammonia gas on inorganic and organic sorbents at 298.15 K. Journal of Chemical & Engineering Data. 2001;**46**:391-399. DOI: 10.1021/je000273+

[61] Huang N, Ding XS, Kim J, Ihee H, Jiang DL. A photoresponsive smart covalent organic framework. Angewandte Chemie, International Edition. 2015;**54**:8704-8707. DOI: 10.1002/ange.201503902

[62] Crowe JW, Baldwin LA, McGrier PL. Luminescent covalent organic frameworks containing a homogeneous and heterogeneous distribution of dehydrobenzoannulene vertex units. Journal of the American Chemical Society. 2016;**138**:10120-10123. DOI: 10.1021/jacs.6b06546

[63] Wan S, Guo J, Kim J, Ihee H, Jiang DL. A belt-shaped, blue luminescent, and semiconducting covalent organic framework. Angewandte Chemie, International Edition. 2008;**47**:8826-8830. DOI: 10.1002/anie.200803826

[64] Hu H, Yan Q, Ge R, Gao YA. Covalent organic frameworks as heterogeneous catalysts. Chinese Journal of Catalysis. 2018;**39**:1167-1179. DOI: 10.1016/S1872-2067(18)635057-8d

[65] Ma D, Wang Y, Liu AA, Li SH, Lu CC, Chen CC. Covalent organic frameworks: Promising materials as heterogeneous catalysts for C-C bond formations. Catalysts. 2018;**8**:404. DOI: 10.3390/catal8090404

[66] Ding SY, Gao J, Wang Q, Zhang Y, Song WG, Su CY, et al. Construction of covalent organic framework for catalysis: Pd/COF-LZU1 in Suzuki-miyaura coupling reaction. Journal of the American Chemical Society. 2011;**133**:19816-19822. DOI: 10.1021 /ja 206 846p

[67] Mu MM, Wang YW, Qin YT, Yan XL, Li Y, Chen LG. Two-dimensional imine-linked covalent organic frameworks as a platform for selective oxidation of olefins. ACS Applied Materials & Interfaces. 2017;**9**: 22856-22863. DOI: 10.1021/acsami.7b05870

[68] Zhang JQ, Peng YS, Leng WG, Gao YA, Xu FF, Chai JL. Nitrogen ligands in two-dimensional covalent organic frameworks for metal catalysis. Chinese Journal of Catalysis. 2016;**37**:468-475. DOI: 10.1016/S1872-2067(15)61050-6

[69] Hou YX, Zhang XM, Sun JS, Lin S, Qi DD, Hong RR, et al. Good Suzuki-coupling reaction performance of Pd immobilized at the metal-free porphyrin-based covalent organic framework. Microporous and Mesoporous Materials. 2015;**214**:108-114. DOI: 10.1016/j.micromeso.2015.05.002

[70] Zhang WJ, Jiang PP, Wang Y, Zhang J, Gao YX, Zhang PB. Bottom-up approach to engineer a

molybdenum-doped covalent-organic framework catalyst for selective oxidation reaction. RSC Advances. 2014;**4**:51544-51547. DOI: 10.1039/c4ra09304f

[71] Zhao M, Wu CD. Synthesis and postmetalation of a covalent-porphyrinic framework for highly efficient aerobic epoxidation of olefins. Catalysis Communications. 2017;**99**:146-149. DOI: 10.1016/j.catcom.2017.06.001

[72] Xu H, Chen X, Gao J, Lin JB, Addicoat M, Irle S, et al. Catalytic covalent organic frameworks via pore surface engineering. Chemical Communications. 2014;**50**:1292-1294. DOI: 10.1039 /c3cc48813f

[73] Xu H, Gao J, Jiang DL. Stable, crystalline, porous, covalent organic frameworks as a platform for chiral organocatalysts. Nature Chemistry. 2015;7:905-912. DOI: 10.1038/NCHEM.2352

[74] Zhang J, Han X, Wu XW, Liu Y, Cui Y. Multivariate chiral covalent organic frameworks with controlled crystallinity and stability for asymmetric catalysis. Journal of the American Chemical Society. 2017;**139**:8277-8285. DOI: 10.1021/jacs.7b03352

[75] Han X, Xia QC, Huang JJ, Liu Y, Tan CX, Cui Y. Chiral covalent organic frameworks with high chemical stability for heterogeneous asymmetric catalysis. Journal of the American Chemical Society. 2017;**139**:8693-8697. DOI: 10.1021/jacs.7b04008

[76] Han X, Zhang J, Huang JJ, Wu XW, Yuan DQ, Liu Y, et al. Chiral induction in covalent organic frameworks. Nature Communications. 2018;**9**:1294. DOI: 10.1038/s41467-018-03689-9

[77] Dong B, Wang LY, Zhao S, Ge RL, Song XD, Wang Y, et al. Immobilization of ionic liquids to covalent organic frameworks for catalyzing the formylation of amines with CO_2 and phenylsilane. Chemical Communications. 2016;**52**:7082-7085. DOI: 10.1039/c6cc03058k

[78] Xu HS, Ding SY, An WK, Wu H, Wang W. Constructing crystalline covalent organic frameworks from chiral building blocks. Journal of the American Chemical Society. 2016;**138**:11489-11492. DOI: 10.1021/jacs.6b07516

[79] Peng YW, Hu ZG, Gao YJ, Yuan DQ, Kang ZX, Qian YH, et al. Synthesis of a sulfonated two-dimensional covalent organic framework as an efficient solid acid catalyst for biobased chemical conversion. ChemSusChem. 2015;**8**:3208-3212. DOI: 10.1002/cssc.201500755

[80] Wang XR, Han X, Zhang J, Wu XW, Liu Y, Cui Y. Homochiral 2d porous covalent organic frameworks for heterogeneous asymmetric catalysis. Journal of the American Chemical Society. 2016;**138**:12332-12335. DOI: 10.1021/jacs.6b07714

[81] Roeser J, Kailasam K, Thomas P. Covalent triazine frameworks as heterogeneous catalysts for the synthesis of cyclic and linear carbonates from carbon dioxide and epoxides. ChemSusChem. 2012;**5**:1793-1799. DOI: 10.1002/cssc.201200091

[82] Wu Y, Xu H, Chen X, Gao J, Jiang DL. A π-electronic covalent organic framework catalyst: π-walls as catalytic beds for diels-alder reactions under ambient conditions. Chemical Communications. 2015;**51**:10096-10098. DOI: 10.1039/c5cc03457d

[83] Shinde DB, Kandambeth S, Pachfule P, Kumar RR, Banerjee R. Bifunctional covalent organic frameworks with two dimensional organocatalytic micropores. Chemical Communications. 2015;**51**:310-313. DOI: 10.1039/c4cc07104b

[84] Leng WG, Peng YS, Zhang JQ, Lu H, Feng X, Ge RL, et al. Sophisticated design of covalent organic frameworks with controllable bimetallic docking for a cascade reaction. Chemistry--A European Journal. 2016;**22**:9087-9091. DOI: 10.1002/chem.201601334

[85] Qi S, Aguila B, Perman JA, Nguyen TK, Ma SQ. Flexibility matters: Cooperative active sites in covalent organic framework and threaded ionic polymer. Journal of the American Chemical Society. 2016;**138**: 15790-15796. DOI: 10.1021/jacs.6b10629

[86] Pachfule P, Kandambeth S, Díaz DD, Banerjee R. Highly stable covalent organic framework-Au nanoparticles hybrids for enhanced activity for nitrophenol reduction. Chemical Communications. 2014;**50**:3169-3172. DOI: 10.1039/c3cc49176e

[87] Pachfule P, Panda MK, Kandambeth S, Shivaprasad SM, Diaz DD, Banerjee R. Multifunctional and robust covalent organic framework-nanoparticle hybrids. Journal of Materials Chemistry A. 2014;**2**:7944-7952. DOI: 10.1039/c4ta00284a

[88] Bhadra M, Sasmal HS, Basu A, Midya SP, Kandambeth S, Pachfule P, et al. Predesigned metal anchored building block for in situ generation of Pd nanoparticles in porous covalent organic framework: Application in heterogeneous tandem catalysis. ACS Applied Materials & Interfaces. 2017;**9**:13785-13792. DOI: 10.1021/acsami.7b02355

[89] Shi XF, Yao YJ, Xu YL, Liu K, Zhu GS, Chi LF, et al. Imparting catalytic activity to a covalent organic framework material by nanoparticle encapsulation. ACS Applied Materials & Interfaces. 2017;**9**:7481-7488. DOI: 10.1021/acsami.6b16267

[90] Mullangi D, Nandi S, Shalini S, Sreedhala S, Vinod CP, Vaidhyanathan R. Pd loaded amphiphilic COF as catalyst for multi-fold heck reactions, C-C couplings and CO oxidation. Scientific Reports. 2015;**5**:10876. DOI: 10.1038/srep10876

[91] Lu SL, Hu YM, Wan S, Mccaffrey R, Jin YH, Gu HW, et al. Synthesis of ultrafine and highly dispersed metal nanoparticles confined in a thioether-containing covalent organic framework and their catalytic applications. Journal of the American Chemical Society. 2017;**139**:17082-17088. DOI: 10.1021/jacs.7b07918

[92] Fang QR, Gu S, Zheng J, Zhuang ZB, Qiu SL, Yan YS. 3D microporous base-functionalized covalent organic frameworks for size-selective catalysis. Angewandte Chemie, International Edition. 2014;**53**:2878-2882. DOI: 10.1002/anie.201310500

[93] Li H, Pan QY, Ma YC, Guan XY, Xue M, Fang QR, et al. Three-dimensional covalent organic frameworks with dual linkages for bifunctional cascade catalysis. Journal of the American Chemical Society. 2016;**138**: 14783-14788. DOI: 10.1021/jacs.6b09563

[94] Ma YX, Li ZJ, Wei L, Ding SY, Zhang YB, Wang W. A dynamic three-dimensional covalent organic framework. Journal of the American Chemical Society. 2017;**139**:4995-4998. DOI: 10. 1021/jacs.7b01097

Chapter 5

Designed Mesoporous Materials toward Multifunctional Organic Silica Nanocomposites

Hendrik O. Lintang and Leny Yuliati

Abstract

Functionalized mesoporous silica materials (MSMs) using grafting (post-synthesis) and one-pot (co-condensation) synthesis methods of organic functional groups (periodic mesoporous organosilicas, PMOs) have been developed for many emerging applications. To improve the functions, designed MSMs have received particular attention using an organic motif as a molecule of surfactants for the template synthesis with a silica source in the sol-gel reaction. The resulting mesoporous silica materials can provide characteristic multifunctional nanocomposites consisting of a monomer for synthesizing polymer in the silicate nanochannels. Moreover, the nanocomposites can be also synthesized using a self-assembled organic motif for organizing one-dimensional structure in the silicate nanochannels. The resulting hybrid nanomaterials have been mainly reported to provide fluorescent properties. However, the utilization of phosphorescent nanocomposites for specific applications has not yet reported so far. By utilizing a self-assembled metal complex (organometallic), this chapter particularly highlights recent achievements of designed mesoporous silica materials for the fabrication of advanced luminescent nanostructures with phosphorescent properties where the potential applications will be discussed in detail for self-repairing and thermally resistive materials, metal ions sensors, template synthesis nanoparticles, and catalysts. Such better and novel performance can be only achieved using a designed template for the sol-gel synthesis of mesoporous silica nanocomposites.

Keywords: hybrid material, luminescence, mesoporous silica, metal complex, nanocomposite, nanomaterial, one-pot synthesis, organosilica, surfactant, template

1. Introduction

Silica sources have been widely used for the preparation of solid containing pores (porous) materials. According to definition by International Union of Pure and Applied Chemistry (IUPAC) in 1972, porous materials can be defined based on the pore size (diameter). Materials with pore size less than 2 nm, between 2 to 50 nm and 50 to 7500 nm as well as above 7500 nm are classified as microporous, mesoporous, macroporous and megapore materials, respectively [1]. Such porous materials have been widely used for many applications due to their characteristics in pore size and surface area [2].

Japanese and American researchers have independently invented ordered Mesoporous Silica Materials (MSMs) in early 1990 using various kinds of surfactants

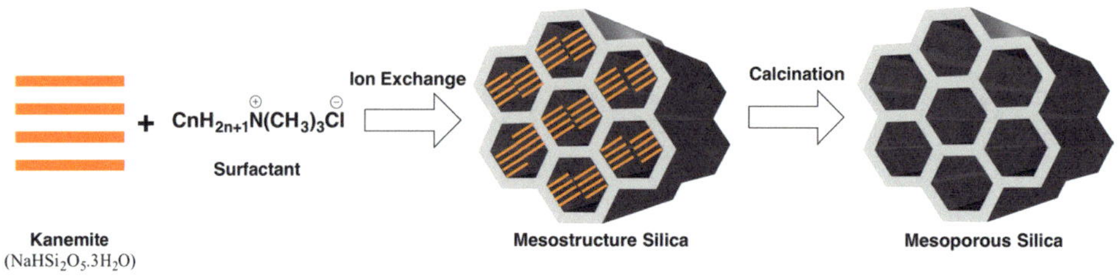

Figure 1.
Synthetic scheme for the formation of Folded Sheet Materials (FSM)–16 through calcination of as-synthesized mesostructure silica from the sol-reaction of kanemite as a silica source and cationic surfactant.

and silica sources. Based on the research by Japanese researchers, the first ordered MSMs have been discovered by using intercalation of cationic micelles in a layered silicate kanemite [3, 4]. These MSMs have a hexagonal array of uniform channels that is called as *Folded Sheet Materials* (FSM)–16. As shown in **Figure 1**, the FSM-16 was prepared from a condensation of three-dimensional silicate networks through sodium (alkali) cation exchange of kanemite with organic cations of alkyltrimethyl-ammonium ions having chloride anion. On the other hands, American researchers from Mobil Research and Development Corporation [5, 6] have also discovered different ordered MSMs using cetyltrimethylammonium bromide (CTAB) as a surfactant for templating silica framework. The resulting MSMs have been classified as *Mobil Composition of Matters* (MCMs)–41 families consisting different geometry of MSMs such as hexagonal (MCM-41), cubic (MCM-48), and lamellar (MCM-50).

Liquid Crystal Templating (LCT) mechanism has been proposed for the formation of the MCMs with a hexagonal structure. In particular, surfactant in their liquid crystal properties (lyotropic) plays an important role for the formation of MCMs [5, 6]. **Figure 2** shows the self-assembly of surfactant in the presence of silica sources to form mesoporous silica. Typically, the surfactant such as CTAB self-assembles to form a micelle and then it will form a rod micelle. Such rod micelle at the certain concentration (*critical micelle concentration*, CMC) can self-assemble to form a structure with a hexagonal geometry. In the presence of organic template, the interaction can occur through oligomerization of silica source to hexagonal packing under acidic or basic condition. It will then produce mesostructure silica containing the surfactant molecules in the silicate nanochannels. By using calcination method, both groups have demonstrated that highly ordered and well-defined nanoscopic channels of mesostructures can be designed and then synthesized with good characteristics such as large surface areas, high thermal and mechanical stability, uniform channel distribution, and pore size modification.

Recent research in the development of ordered MSMs have been widely focused using different kinds of surfactants. For example, *Hexagonal Mesoporous Silica* (HMS) with slightly disordered hexagonal structure and thicker walls as well as superior thermal stability has been synthesized using neutral amine surfactants [7]. By using the same type of surfactant, *Michigan State University* (MSU-1) has been synthesized using neutral polyethylene oxide (PEO) surfactant [8]. Moreover, *Santa Barbara Amorphous* (SBA)-15 has been invented with good characteristic such as large pore size and well tunable in the range of 5–30 nm, thicker pore walls, and two-dimensional (2D) hexagonal structure as well as higher hydrothermal stability. SBA-15 has been synthesized by using amphiphilic triblock-copolymer of poly(ethylene oxide) and poly(propylene oxide) or Pluronic P123 as a surfactant [9].

Such various types of ordered MSMs have been potentially studied for different kinds of applications in catalysts, adsorbents, molecular sieves, drug delivery, sensor, insulating materials and nanometer-scale hosts for optical and electronic materials.

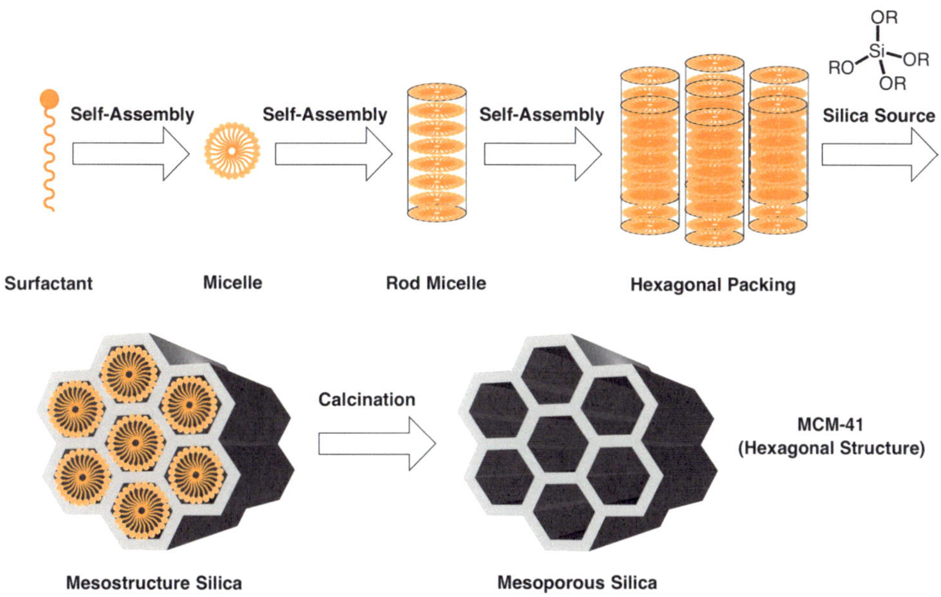

Figure 2.
Synthetic scheme for the formation of Mobil Composition of Matters (MCMs)–41 with a hexagonal from calcination of mesostructure silica from sol-gel reaction of self-assembled surfactant in their micelle form with a silica source.

However, the performance of bulk ordered MSMs have been limited in their applications as mentioned in several review journals [10–14]. Hence, ordered MSMs have been functionalized with organic functional groups as multifunctional nanocomposites.

2. Functionalization of mesoporous silica materials

Functionalization of MSMs with organic functional groups became a major topic of research because it can be used for modifying their chemical properties for specific applications at the surface, wall and/or channel at the nanoscale. Moreover, silica frameworks can be designed with specific properties such as hydrophobicity, hydrophilicity, polarity, catalytic active sites, and optical as well as electronic activities so that the performance can be improved in their applications. In this chapter, the functionalization of MSMs will be reviewed using post-synthetic (grafting) and one-pot (co-condensation) methods [15–20] where short discussion will only be provided consisting of the illustration for the formation of mesoporous silica based on the usage of surfactants. At the end of this subheading, these approaches will be then developed as a template sol-gel synthesis for designed multifunctional MSMs where organic chemists can design the desired functions of MSMs for better performance.

2.1 Functionalization with post-synthetic method

Post-synthetic method has been early proposed for functionalization of ordered MSMs by grafting silanol groups consisting of functional organic moieties (**F** star) at the surface of ordered MSMs. **Figure 3** shows the method where the functional groups in organosilanes can form a covalent bonding with silanol of the MSMs. This method has been firstly introduced from the early discovery of MCM-41 [6] as well as kanemite [3] using trimethylsilyl chloride ($[(CH_3)_3SiCl]$) and hexamethyldisiloxane ($[(CH_3)_3Si)_2O]$).

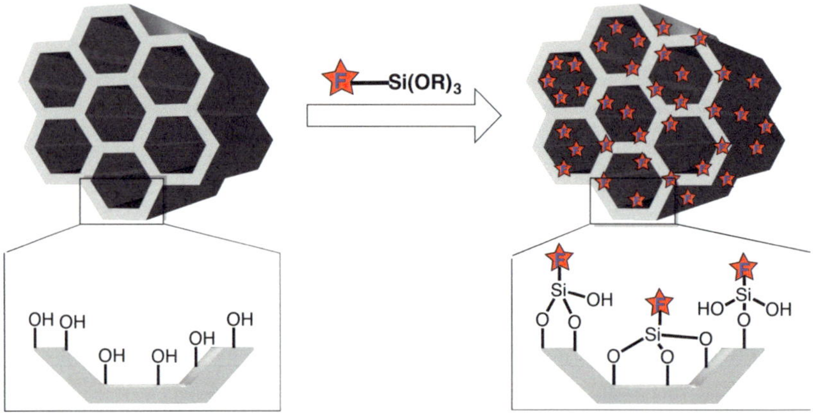

Figure 3.
Post-synthetic method for the functionalization of ordered MSMs.

In the post-synthetic methods, three main approaches have been reported so far using organosilanes and their further modification. Such approaches can be described as follows [17, 21]:

a. First approach using modification of silica surface with organosilanes such as alkoxysilanes [22, 23], chlorosilanes [6, 24, 25], and silylamines as well as disilazane [26].

b. Second approach using modifications of previously grafted surface under procedures (a) with other functional groups.

c. Third approach using transformation of grafted surface under procedures (a) and (b) with additional treatments.

The above approaches can be categorized as grafting with passive (low reactivity) and active (high reactivity) surface groups [16]. However, pore blocking and non-homogenous distribution of organosilanes in the silicate nanochannels are the main problems for the improvement of the properties and performance [19]. Therefore, the functionalized MSMs have been mainly developed for applications in the removal of metal ions and adsorbents [16].

2.2 Functionalization with co-condensation method

As shown in **Figure 4**, co-condensation method has been used for the functionalization of ordered MSMs by mixing organosilanes with silica source and surfactant or structure-directing agent during the sol-gel synthesis. In this method, covalent bonding of a nonhydrolyzable Si—C bond in organosilanes will be hydrolyzed in the sol-gel reaction to form a silica network of mesostructure silica with functional groups in the walls and/or pores [16–19]. Generally, these functional materials have been mainly used for development of optical nanocomposites where specific organic chromophores with such properties can be encapsulated [27].

Two main approaches have been used for co-condensation method based on the types of organosilanes such as mono and bridge silanes attached in the organic functional groups:

2.2.1 Co-condensation method with mono-organosilanes

Co-condensation method with mono-organosilanes has been carried out using mono-functionalization of organosilane as firstly proposed with a simple structure.

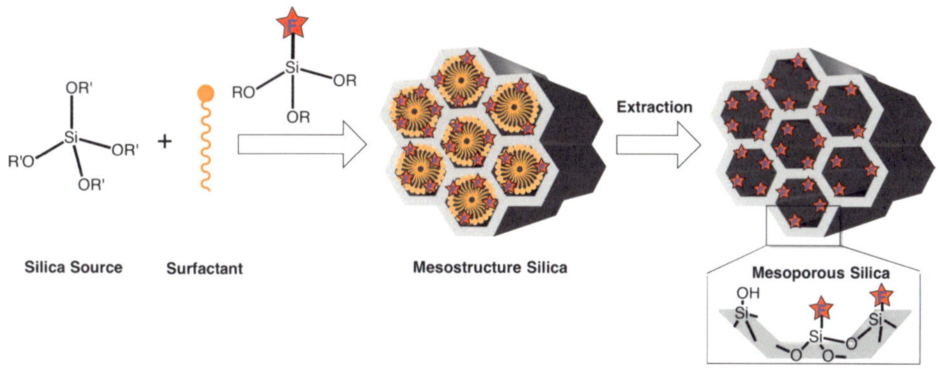

Figure 4.
Co-condensation method for the organic modification of ordered MSMs with organosilanes.

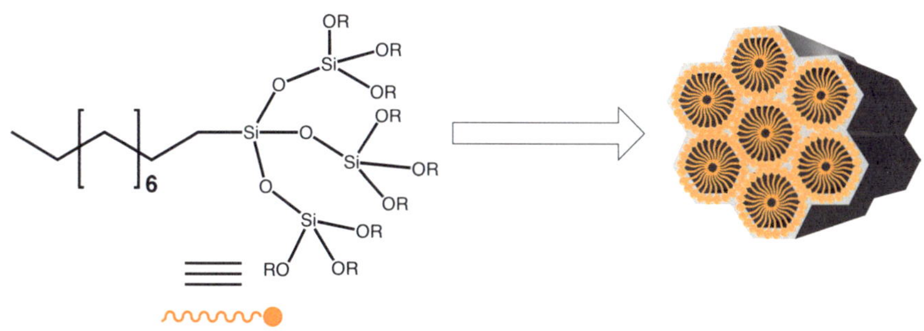

Figure 5.
Formation of ordered MSMs from siloxane-based dendritic bearing long alkyl chain as a surfactant and a silica source in the sol-gel synthesis of MSMs.

For instance, phenyltriethoxysilane (PTES) or n-octyltriethoxysilane (OTES) [28], 3-aminopropy1trimethoxysilane (APTMS) and 2-cyanoethyltrimethoxysilane (CETMS) [29] and vinyltriethoxysilane (VTES) [30] have been used as mono-organosilanes. However, the main problems in this method have been found such as only 40% of maximum functionalization, reduction in pore diameter and volume as well as decreasing of specific surface area and difficulty in removing of surfactants.

Recently, synthesis of ordered MSMs with mono-organosilanes as both a silica source and a surfactant or structural-directing agent was reported using n-tetradecyldimethy(3-trimethoxysilylpropyl) ammonium chloride [31]. As shown in **Figure 5**, Shimojima and Kuroda in 2003 [32] have initiated the use of more complicated organosilane structure such as dendritic type of alkoxysilyl-terminated alkyloligosiloxane [32]. One year later, using dendrimer carbosilanes with its first and second generations, Landskron and Ozin then modified such more complicated structure of the dendritic organosilanes to prepare "Periodic Mesoporous Dendrisilicas" (PMD) as reported in Science 2004 [33]. However, the functional groups attached to the organosilanes in ordered MSM have not yet been applied for specific functions due to its low functionality in the silica framework.

Zhang *et al*. in 2004 [34] was introduced another approach to improve the function of MSMs even the functional surfactant has rigid structure. In this case, they have introduced an amphiphilic alanine-containing alkoxysilane with a cleavable alkyl chain at the ester group and condensable cationic head-group at siloxane (Lizard template) as shown in **Figure 6**. When they performed acid hydrolysis to the template, it can provide alanine functional groups consisting of cationic ammonium head-group with silanol. Interestingly, alanine functional group in this hybrid material can be used as a heterogeneous catalyst for the acetalization of

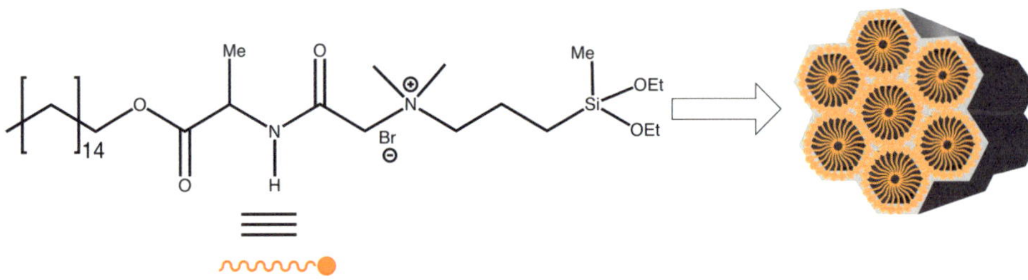

Figure 6.
Formation of ordered MSM from a "lizard" template through sol-gel synthesis and followed by acid hydrolysis.

cyclohexanone in ethanol with high conversion [35]. Such performance was proposed based on the formation of hydrogen bonding from the presence of cyclohexanone and ethanol at the hydrophobic domain at the inner shell and hydrophilic domain at the outer shell with core-shell interface containing amide and carbonyl groups.

2.2.2 Co-condensation method with bridged-organosilanes

Co-condensation method with bridged-organosilanes has been carried out using two trialkoxysilyl groups connected by a functional organic bridge, called as bis-organosilanes to produce "Periodic Mesoporous Organosilicas" (PMOs) as invented independently in 1999 by three research groups [15, 36, 37]. In contrast to above methods, the organic functional groups are incorporated in the three-dimensional network structure of the silica wall through two covalent bonds as shown in **Figure** 7. They are many organic functional groups have been used as bis-organosilanes where the **F** star are methane, ethane, benzene, stilbene, azobenzene, phenyl vinylene, cyclic ethane, dendritic ethane, *etc*. and then used for the synthesis of MSMs with homogeneous distribution of the functional groups in the wall.

Inagaki *et al.* in 2002 [38] reported one of good example in the synthesis of PMO. They have used 1,4-bis(triethoxysilyl)-benzene (BTEB) where the organosilane is based on benzene structure as the functional group. After reaction in a basic condition using sodium hydroxide for 20 hours, the mesostructure silica was successfully synthesized as a white precipitate, and was extracted to remove the surfactant to give plate-like particles of PMO. It shall be noted that PMO has found to form an alternately structure from the layers of hydrophilic silicate and hydrophobic benzene.

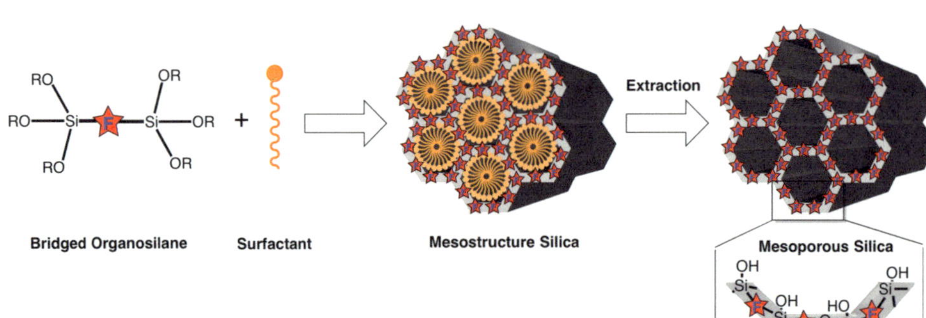

Figure 7.
Co-condensation method for the synthesis of ordered periodic Mesoporous Organosilicas (PMOs) with bridged-organosilanes and surfactant.

2.3 Functionalization with template sol: Gel synthetic method

Since PMOs have been only developed to functionalize the pore wall of MSMs, it is really interesting to guarantee the presence of organic functional groups in the silicate nanochannels. Moreover, the channels of the silicate framework shall be fully occupied by the organic moieties so that they are many functional sites for specific applications. Toward this expectation, it is necessary to design the surfactant containing of functional organic moieties. Therefore, templated sol-gel synthesis method has been proposed using a surfactant in the functionalization of MSMs to form desired multifunctional nanocomposites. This approach has been proposed by using surfactant bearing organic functional groups (called as a "functional surfactant") as a template in the sol-gel synthesis of ordered MSMs.

Three research groups have firstly proposed in the synthesis of ordered MSM by functional surfactants using amphiphilic diacetylenic monomers [39, 40] and an amphiphilic phthalocyanine [41]). In particular, the functional organic groups will covalently bond in the surfactant molecules where it can be prepared in few step organic reactions. By using this functional surfactant as shown in **Figure 8**, the resulting functional MSMs in the sol-gel synthesis will guarantee dense filling of the silicate nanochannels with the acetylenic functional groups [39, 40]. Therefore, based on the desired functions, the researcher shall decide the organic moieties to be attached covalently and number of functional groups so that the retro-synthesis shall be determined for the synthesis of that surfactants.

In this chapter, the template sol-gel synthesis method will be discussed based on the polymerizable and self-assembled functional surfactants:

2.3.1 Polymerizable surfactant

Polymerizable surfactant as a template in the sol-gel synthesis of ordered MSMs has been independently introduced in 2001 using diacetylene monomers [39, 40]. **Figure 9** shows the example using cationic type of functional surfactants as reported by Aida and Tajima [39]. Polymerization of the resulting mesostructured materials containing dense filling of diacetylenic monomers by photo or thermal treatments gave polydiacetylenes in the nanoscopic channels. In particular, these polydiacetylene nanocomposites have showed photoluminescent hybrid materials with elongated effective conjugation compared to its polymer from monomer directly, indicating the confinement effect in the nanoscale.

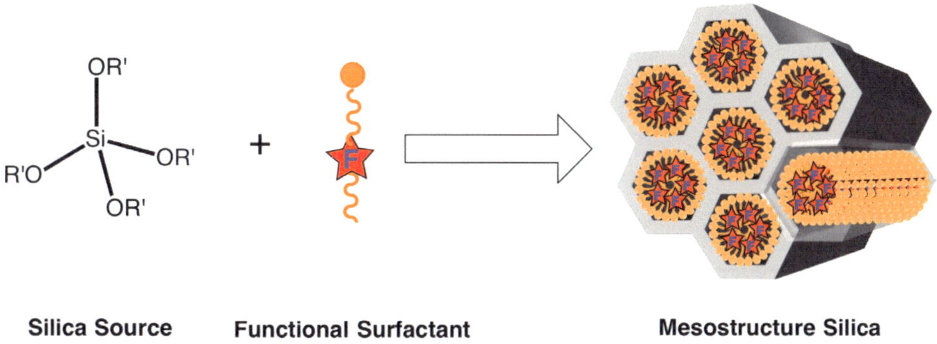

Figure 8.
Co-condensation of ordered MSMs using a surfactant bearing organic functional group as a template for the sol-gel synthesis with silica source.

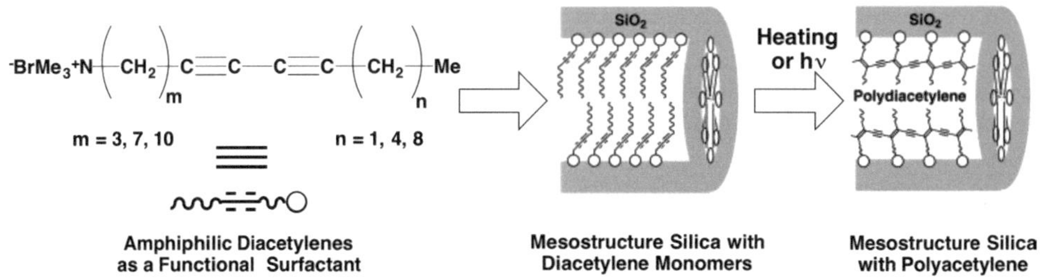

Figure 9.
Sol-gel synthesis of ordered MSMs using cationic surfactants bearing organic functional group of diacetylene monomers as a functional template.

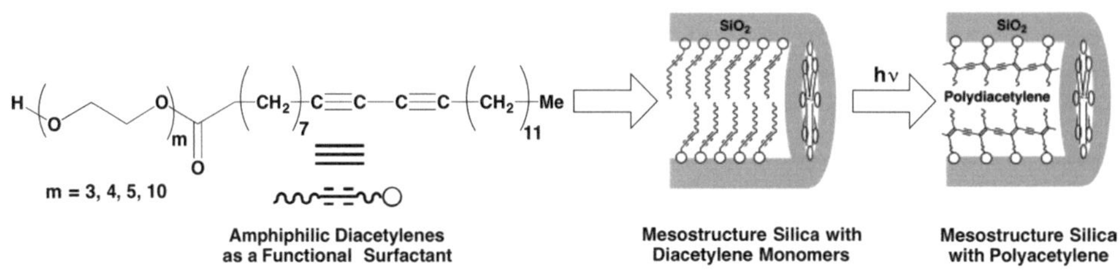

Figure 10.
Sol-gel synthesis of ordered MSMs using amphiphilic diacetylenic monomers surfactants as a functional template.

Figure 10 shows the example using amphiphilic type of functional surfactants as reported by Lu *et al.* [40]. Interestingly, they have exhibited chromatic of the blue nanocomposite films to red emission. Such changes have showed not only solvatochromic properties upon exposure to polar solvents to diffuse to hydrophilic ethylene glycol parts, but also thermochromic properties upon heating to form interaction between silanol moieties and alkyl side-chains via hydrogen bonding, as well as mechanochromic properties upon abrasion to form mechanical damage.

In 2003, pyrrole monomers having C10 and C12 alkyl chains were used as a template to synthesize mesoporous silica with both a hexagonal and lamellar geometries. The ordered MSMs with dense filling of pyrrole monomers could be oxidatively polymerized by simply dipping the spin-coated film into initiator $FeCl_3$ (2 M) in diethyl ether for 1 min [42]. The resulting polypyrrole chains insulated in the hexagonal silicate nanochannels significantly suppressed recombination of polarons into bipolarons, while the lamellar silicate nanochannels afforded spatial freedom for the electron recombination. In the same year, thiophene was used as a template for the polymerization in mesoporous silica to give photoluminescent silica nanocomposite with an alignment of conjugated 100 nm polymer chains in the channels [43].

Other functional polymer nanocomposites have received great attention for the development new materials with good characteristics. To work in this field, redox active functional groups using styrene as a monomer and template was successfully prepared as mesostructured silica/polystyrene nanocomposite film with characteristic of charge transport through a diffusion process [44]. In other reports, luminescent polymers were also successfully prepared using phenylene ethylene [45] and thiophene [46] as a template to give poly(phenylene ethylene) and poly(thiophene) where the nanocomposites showed fluorescent properties with good properties such as higher anisotropic degree and tunable color emission.

2.3.2 Self-assembled surfactant

Self-assembled surfactant as a functional template in the sol-gel synthesis of ordered MSMs has been firstly introduced using an amphiphilic copper phthalocyanine as shown in **Figure 11** [41]. To synthesize the surfactant, the eight decoxy of hydrophobic parts containing triethylene glycol of hydrophilic part was synthesized as amphiphilic side chains in the alkyl bromide. Reaction with copper(I) cyanide through Williamson ether substitution reaction, amphiphilic copper phthalocyanine can be synthesized. This surfactant will form disk-shape (discotic) for self-assembly via noncovalent π–π stacking interactions so that 1D molecular assembly can be insulated within the channels of mesoporous silica.

The columnar assembly in the channels of ordered MSMs can be also synthesized by using charge-transfer interactions of amphiphilic triphenylene as a donor and various acceptors with 1 to 1 mole ratio [47]. In this case, the charge-transfer adduct shall able to form columnar assembly where the donor and acceptor molecules shall be able to be alternated. Without the acceptor molecule, less ordered MSMs have been the amphiphilic triphenylene as a donor molecule prepared. By selecting specific acceptor molecules to be paired, the resulting transparent nanocomposite films can provide color-tunable materials with blue to red with peak top of absorption at 490 nm for 2,4,7-trinitro-9-fluorenone (TNF), 548 nm for 1,2,4,5-tetracyanobenzene (TCNB), 615 nm for 2,3,6,7,10,11-hexacyano-hexaazatriphenylene (HAT), 700 nm for chloranil (CA) and 890 nm for, 7,7,8,8-tetracyanoquinodimethane (TCNQ).

Since less ordered MSMs was found when triphenylene as a donor and functional surfactant was only used in the sol-gel synthesis without the presence of acceptor molecules, the new strategy to synthesize high ordered MSMs is required. By using ethanol vapor in the closed system to the as-synthesized mesostructured silica, the same triphenylene was immersed to ethanol vapor in one drop to give high ordered MSMs in 3 hours at room temperature [48]. The ethanol vapors have a function to increase structure regularity as shown from increasing intensity of the main diffraction peak at d_{100}. Another approach was reported by using the hydrophilic version containing of 18 ethylene glycol side-chains of HBC for the synthesis of nanocomposites [49]. The HBC was firstly prepared using $Pd(PPh_3)_4$ in CuI as catalyst for coupling reaction with hexaiodo-*peri*-hexabenzo-coronene in the presence of Pd catalyst. Such high hydrophilicity HBC was only produced MSMs with less ordered structure as shown from less intense of the main diffraction peak at d_{100}. Hence, it is really necessary to balance the amphiphilicity from the presence of hydrophobic and hydrophilic side chains of the functional surfactant when there is only one molecule as a template.

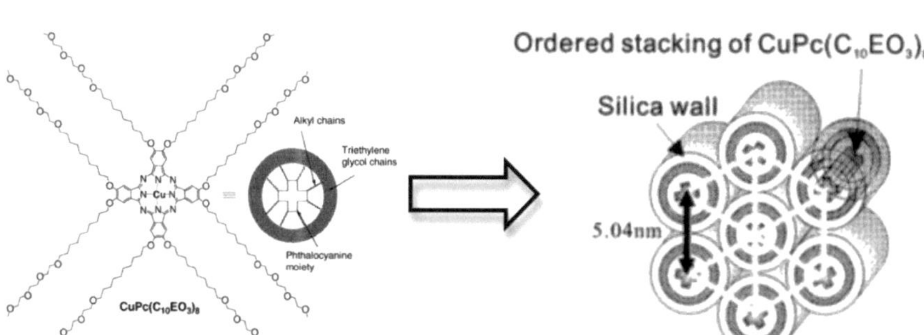

Figure 11.
Sol-gel synthesis of ordered MSMs using amphiphilic copper phthalocyanine as a template.

Luminescent nanomaterials with phosphorescent properties have received particular attention for the development of optical materials. To this aims, a weak interaction of metal–metal bonding can be also used to construct columnar assembly with luminescent property. Otani *et al.* [50] was reported that an amphiphilic bispyridine platinum(II) complex, as synthesized from 2,2-bipyridine and K_2PtCl_4, can be used as a functional surfactant where Pt double-salt from linear chain of Pt-Pt metallophilic interactions between Pt(II) complex with another Pt(II) salt can self-assemble with silica source to form luminescent nanocomposites. Interestingly, these nanocomposites showed phosphorescent properties with an emission peak centered at 619 nm upon an excitation at 495 nm.

Recently, trinuclear gold(I) pyrazolate complex ([Au_3Pz_3]) bearing amphiphilic side chains was successfully synthesized to be used as a functional surfactant. This complex showed not only liquid crystalline properties with mesophase range over a wide range of room temperature, but also phosphorescent properties around 700 nm with red emission from Au(I)-Au(I) interactions. Moreover, this surfactant can be utilized as a template for synthesizing transparent mesoporous silica nanocomposite films. The resulting mesoporous silica film ([Au_3Pz_3]/$silica_{hex}$) revealed not only phosphorescent properties of red light at (693 nm with the same excitation wavelength) and longer lifetime (7.8 μs) but also perfect self-repairing with stepwise heating until 140°C in the nanoscale [51]. Such performance can be achieved due to a nanoscopic template effect where the ethylene glycol parts at the side chains anchored in the silica wall play an important role to induce the self-repairing of the pyrazole core on cooling at room temperature thermodynamically. In higher temperature under kinetic control, it is impossible to be achieved due to the strong interaction of the paraffinic side chains. In contrast, such self-repairing in the nanoscale from the light-emitting capability of highly phosphorescent nanocomposites with a hexagonal structure cannot be observed with its lamellar geometry [Au_3Pz_3]/$silica_{lam}$ and bulk Au_3PZ_3.

In the development of thermally resistive nanomaterials, by using a thin film [Au_3Pz_3]/$silica_{hex}$, the phosphorescent complex in the nanochannels was treated with thermal treatment at the elevated temperature with stepwise or direct heating. Based on the phenomena of thermal quenching, the nanocomposites showed self-repairing capability at room temperature autonomously when was directly heated to 160°C with a heating time in 11 min (luminescent intensity in 59%) compared to stepwise heating in 52 min for every measurement in every 10°C (luminescent intensity in 20%). Such performance can be supported by the presence of endothermic peak as shown from differential scanning calorimetry (DSC) thermogram at 164°C (first heating) or 158°C (second heating). Both peaks can be assigned as a melting temperature from the deformation of metallophilic interaction in the columnar assembly. Hence, self-healing capability in these nanocomposites can be designed based on the heating treatments where perfect autonomously restoration can be achieved with stepwise heating (20–140°C) [51] or direct heating (20–160°C) [52]. Moreover, such capability can be also observed when the stepwise heating was performed until 200°C [53]. All of these phenomena can be monitored by not only the emission changes but also color changes. Hence, this performance can be potentially utilized as temperature sensor with imaging of the changes in intensity and color emission [54]. Such performance cannot be achieved for bulk luminescent compounds or materials with a lamellar structure where thermal quenching can be easily occurred to reduce the light-emitting capability [51–54].

Metal ion sensors have also received deep attention in the research of nanomaterials. In our study, thin film of this phosphorescent mesoporous silica nanocomposite was applied as a chemosensor of silver ions by simply dipping the thin film into the solution of silver triflate (10–100 μM of AgOTf). The transparent thin film

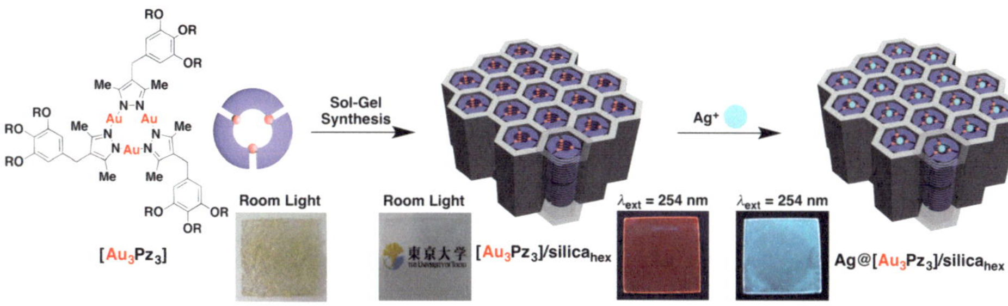

Figure 12.
Sol-gel synthesis of ordered MSMs $[Au_3Pz_3]$/silaca$_{hex}$ as a transparent thin film from the pale yellow of sticky amphiphilic trinuclear gold(I) pyrazolate complex $[Au_3Pz_3]$ as a template and then the utilization for the detection of Ag^+ ion to form Ag@$[Au_3Pz_3]$/silca$_{hex}$.

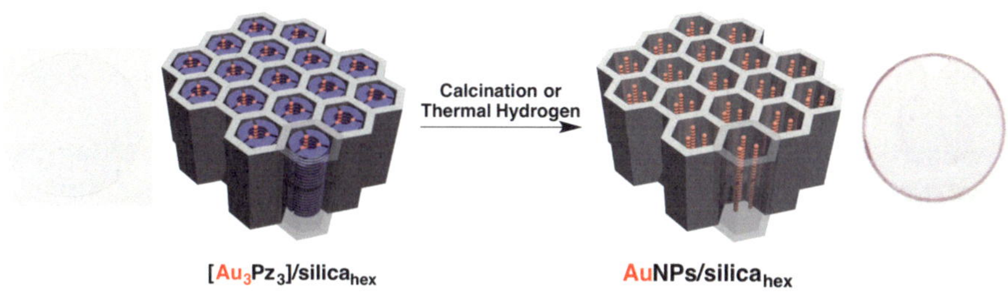

Figure 13.
Thermal treatments for the synthesis of AuNPs in the silicate nanochannels.

$[Au_3Pz_3]$/silica$_{hex}$ with red emission of Au-Au interaction at 693 nm was changed to green (486 nm) of Au-Ag interaction. Moreover, the XPS study with depth profiling indicates that Ag^+ ions were permeated into the silicate nanochannels where the original red emission can be simply recovered in chloroform using cetyltrimethylammonium chloride with the same concentration [55]. **Figure 12** shows the above phenomena from the synthesis of amphiphilic surfactant $[Au_3Pz_3]$ as a template for the sol-gel synthesis of mesoporous silica nanocomposite $[Au_3Pz_3]$/silica$_{hex}$ as a thin film and then was used for sensing Ag^+ ions with color changes from red to green.

Gold nanoparticles (**AuNPs**) have also received great attention in many applications. In our research group, the resulting $[Au_3Pz_3]$/silica$_{hex}$ can be used as a source to form AuNPs in the silicate nanochannels as shown in **Figure 13**. It is important to be noticed that the columnar assembly of $[Au_3Pz_3]$ was fully occupied the silicate channels so that it can provide dense filling of AuNPs in the pore. By using the heat treatment with calcination until 450°C for 3 hours, we have found that the transparent thin film $[Au_3Pz_3]$/silica$_{hex}$ was changed pink thin film, indicating the formation of **AuNP**/silica$_{hex}$. The formation of AuNPs were confirmed by using TEM with the appearance of spherical particles, XRD with the presence of a diffraction peak at 38.2° and UV-Vis spectrometer with the presence of surface plasmon resonance (SPR) peak at 544 nm [56]. Moreover, by using thermal hydrogen treatment at 250°C for hours, it provided the purplish-pink thin film with SPR peak at 558 nm due to the decreasing of particle size [57]. On the other hands, the main diffraction peak at the small-angle area for indicating the hexagonal structure were shifted to higher angle due to the formation of more higher order structure with the same geometry and increasing in pore size. Such template synthesis of nanoparticles can be performed with the confinement effect of gold complexes in the silicate nanochannels. Further research on the detail investigation of the effect on temperature during the heat treatments and applications of **AuNPs** nanocomposites such as a catalyst will be the interesting reports to be discussed.

3. Conclusions

It can be concluded that ordered MSMs have been organically functionalized with post-synthetic grafting and co-condensation methods with organosilanes. However, these methods gave many limitations in their applications. To solve these issues, new functionalization approach with amphiphilic polymerizable monomer and self-assembled discotic surfactants, called as a functional surfactant, have been used as both a template and an attached-organic functional group based on the desired function in the sol-gel synthesis method. The resulting mesoporous silica nanocomposites in the powder or thin film have gave excellent properties compared to the bulk form. Since noncovalent interactions have received many attentions in the development of soft matters, the utilization of this weak π–π and metal–metal bondings as a discotic surfactant was utilized for templating mesoporous silica nanocomposites. By using phosphorescent metal complex, mesoporous silica/gold complex nanocomposite with a hexagonal structure was successfully fabricated as a thin film with good properties such as self-repairing in the nanoscale, thermally resistive and imaging materials, metal ion sensor, and template synthesis of nanoparticles. Toward these applications, it is possible to use the resulting gold nanoparticles as a heterogeneous catalyst in catalytic reactions. Another outlook of these designed nanocomposites is the potential to engineer the silica wall so that the channel and wall can simultaneously work to give synergistic effect on their applications. Moreover, ordered MSMs with chiral and photochromic columnar assemblies with combination of discotic surfactants and condensable siloxanes will be also interesting perspective for the development of memory nanomaterials.

Acknowledgements

The authors thank Directorate General of Strengthening Research and Development, Ministry of Research, Technology and Higher Education of The Republic of Indonesia for the financial support through Fundamental Research (Penelitian Dasar) Grant. The authors also acknowledge Universitas Ma Chung for the financial support.

Author details

Hendrik O. Lintang[1,2]* and Leny Yuliati[1,2]

1 Department of Chemistry, Faculty of Science and Technology, Universitas Ma Chung, Malang, Indonesia

2 Ma Chung Research Center for Photosynthetic Pigments (MRCPP), Universitas Ma Chung, Malang, Indonesia

*Address all correspondence to: hendrik.lintang@machung.ac.id

References

[1] IUPAC. Manual of symbols and terminology, Appendix 2, Pt. 1. Colloid and surface chemistry. Pure and Applied Chemistry. 1972;**31**:578

[2] Ishizaki K, Komarneni S, Nanko M, editors. Porous Materials: Process Technology and Applications. 1st ed. Netherlands: Kluwer Academic Publisher; 1998. p. 234

[3] Yanagisawa T, Shimizu T, Kuroda K, Kato C. The preparation of alkyltrimethylammonium-kanemite complexes and their conversion to mesoporous materials. Bulletin of the Chemical Society of Japan. 1990;**63**: 988-992. DOI: 10.1246/bcsj.63.988

[4] Inagaki S, Fukushima Y, Kuroda K. Synthesis of highly ordered mesoporous materials from a layered polysilicate. Journal of the Chemical Society, Chemical Communications. 1993:680-682. DOI: 10.1039/C39930000680

[5] Kresge CT, Leonowicz ME, Roth WJ, Vartuli JC, Beck JS. Ordered mesoporous molecular sieves synthesized by a liquid-crystal template mechanism. Nature. 1992;**359**:710-712. DOI: 10.1038/359710a0

[6] Beck JS, Vartuli JC, Roth WJ, Leonowicz ME, Kresge CT, Schmitt KD, et al. A new family of mesoporous molecular sieves prepared with liquid crystal templates. Journal of the American Chemical Society. 1992;**114**:10835-10843. DOI: 10.1021/ja00053a020

[7] Tanev PT, Pinnavaia TJ. A neutral templating route to mesoporous molecular sieves. Science. 1995;**267**:865-867. DOI: 10.1126/science.267.5199.865

[8] Bagshaw SA, Prouzet E, Pinnavaia TJ. Templating mesoporous molecular sieves by nonionic polyethylene oxide surfactants. Science. 1995;**269**:1242-1244. DOI: 10.1126/science.269.5228.1242

[9] Zhao D, Feng J, Huo Q, Melosh N, Fredrickson GH, Chmelka BF, et al. Triblock copolymer syntheses of mesoporous silica with periodic 50 to 300 angstrom pores. Science. 1998;**279**:548-552. DOI: 10.1126/science.279.5350.548

[10] Zhao XS, Max Lu GQ, Millar GJ. Advances in mesoporous molecular sieve MCM-41. Industrial & Engineering Chemistry Research. 1996;**35**:2075-2090. DOI: 10.1021/ie950702a

[11] Ying JY, Mehnert CP, Wong MS. Synthesis and applications of supramolecular-templated mesoporous materials. Angewandte Chemie International Edition. 1999;**38**:56-77. DOI: 10.1002/(SICI)1521-3773(19990115)38:1/2<56:AID-ANIE56>3.0.CO;2-E

[12] Scott BJ, Wirnsberger G, Stucky GD. Mesoporous and mesostructured materials for optical applications. Chemistry of Materials. 2001;**13**:3140-3150

[13] Wan Y, Shi Y, Zhao D. Designed synthesis of mesoporous solids via nonionic-surfactant-templating approach. Chemical Communications. 2009:897-926. DOI: 10.1039/B610570J

[14] Kresge CT, Roth W. The discovery of mesoporous molecular sieves from the twenty year perspective. Chemical Society Reviews. 2013;**42**:3663-3670. DOI: 10.1039/C3CS60016E

[15] Asefa T, MacLachlan MJ, Coomb N, Ozin GA. Periodic mesoporous organosilicas with organic groups inside the channel walls. Nature. 1999;**402**:867-871. DOI: 10.1038/47229

[16] Stein A, Melde BJ, Schroden RC. Hybrid inorganic-organic mesoporous silicates-nanoscopic reactors coming of age. Advanced Materials. 2000;**12**:1403-1419. DOI: 10.1002/1521-4095(200010)12:19<1403::AID-ADMA1403>3.0.CO;2-X

[17] Sayari A, Hamoudi S. Periodic mesoporous silica-based organic-inorganic nanocomposite materials. Chemistry of Materials. 2001;**13**:1731-1736

[18] Hatton B, Landskron K, Whitnall W, Perovic D, Ozin GA. Past, present, and future of periodic mesoporous organosilicas–the PMOs. Accounts of Chemical Research. 1992;**38**:305-312. DOI: 10.1021/ar040164a

[19] Hoffmann F, Cornelius M, Morell J, Fröba M. Silica-based mesoporous organic–inorganic hybrid materials. Angewandte Chemie International Edition. 2006;**45**:3216-3251. DOI: 10.1002/anie.200503075

[20] Ariga K, Vinu A, Yamaguchi Y, Ji Q, Hill JP. Nanoarchitectonics for mesoporous materials. Bulletin of the Chemical Society of Japan. 2012;**85**:1-32. DOI: 10.1246/bcsj.20110162

[21] Maschmeyer T. Derivatised mesoporous solids. Current Opinion in Solid State & Materials Science. 1998;**3**:71-78. DOI: 10.1016/S1359-0286(98)80068-5

[22] Sutra P, Brunel D. Preparation of MCM-41 type silica-bound manganese(Il1) Schiff-base complexes. Chemical Communications. 1996:2485-2486. DOI: 10.1039/CC9960002485

[23] Diaz JF, Balkus KJ Jr, Bedioui F, Kurshev V, Kevan L. Synthesis and characterization of cobalt-complex functionalized MCM-41. Chemistry of Materials. 1997;**9**:61-67. DOI: 10.1021/cm960228e

[24] Chen J, Li Q, Xu R, Xiao F. Distinguishing the silanol groups in the mesoporous molecular sieve MCM-41. Angewandte Chemie International Edition in English. 1995;**34**:2694-2696. DOI: 10.1002/anie.199526941

[25] Zhao XS, Lu GQ. Modification of MCM-41 by surface silylation with trimethylchlorosilane and adsorption study. The Journal of Physical Chemistry B. 1998;**102**:1556-1561. DOI: 10.1021/jp972788m

[26] Anwander R, Nagl I, Widenmeyer M, Engelhardt G, Groeger O, Palm C, et al. Surface characterization and functionalization of MCM-41 silicas via silazane silylation. The Journal of Physical Chemistry B. 2000;**104**:3532-3544. DOI: 10.1021/jp993108d

[27] Nicole L, Boissiere C, Grosso D, Quach A, Sanchez C. Mesostructured hybrid organic–inorganic thin films. Journal of Materials Chemistry. 2005;**15**:3598-3627. DOI: 10.1039/B506072A

[28] Burkett SL, Sims SD, Mann S. Synthesis of hybrid inorganic-organic mesoporous silica by co-condensation of siloxane and organosiloxane precursors. Chemical Communications. 1996;**0**:1367-1368. DOI: 10.1039/CC9960001367

[29] Macquarrie DJ. Direct preparation of organically modified MCM-type materials. Preparation and characterization of aminopropyl-MCM and 2-cyanoethyl-MCM. Chemical Communications. 1996:**0**;1961-1962. DOI: 10.1039/CC9960001961

[30] Lim MH, Blanford CF, Stein A. Synthesis and characterization of a reactive vinyl-functionalized MCM-41: Probing the internal pore structure by a bromination reaction. Journal of the American Chemical Society. 1997;**119**:4090-4091. DOI: 10.1021/ja9638824

[31] Huo Q, Margolese DI, Stucky GD. Surfactant control of phases in the synthesis of mesoporous silica-based

materials. Chemistry of Materials. 1996;**8**:1147-1160. DOI: 10.1021/cm960137h

[32] Shimojima A, Kuroda K. Direct formation of mesostructured silica-based hybrids from novel siloxane oligomers with long alkyl chains. Angewandte Chemie International Edition. 2003;**42**:4057-4060. DOI: 10.1002/anie.200351419

[33] Landskron K, Ozin GA. Periodic mesoporous dendrisilicas. Science. 2004;**306**:1529-1532. DOI: 10.1126/science.1104555

[34] Zhang Q, Ariga K, Okabe A, Aida T. A condensable amphiphile with a cleavable tail as a "lizard" template for the sol-gel synthesis of functionalized mesoporous silica. Journal of the American Chemical Society. 2004;**126**:988-989. DOI: 10.1021/ja039515r

[35] Otani W, Kinbara K, Zhang Q, Ariga K, Aida T. Catalysis of a peptidic micellar assembly covalently immobilized within mesoporous silica channels: Importance of amphiphilic spatial design. Chemistry—a European Journal. 2007;**13**:1731-1736. DOI: 10.1002/chem.200601099

[36] Inagaki S, Guan S, Fukushima Y, Ohsuna T, Terasaki O. Novel mesoporous materials with a uniform distribution of organic groups and inorganic oxide in their frameworks. Journal of the American Chemical Society. 1999;**121**:9611-9614. DOI: 10.1021/ja9916658

[37] Melde BJ, Holland BT, Blanford CF, Stein A. Mesoporous sieves with unified hybrid inorganic/organic frameworks. Chemistry of Materials. 1999;**11**:3302-3308. DOI: 10.1021/cm9903935

[38] Inagaki S, Guan S, Ohsuna T, Terasaki O. An ordered mesoporous organosilica hybrid material with a crystal-like wall structure. Nature. 2002;**416**:304-307. DOI: 10.1038/416304a

[39] Aida T, Tajima K. Photoluminescent silicate microsticks containing aligned nanodomains of conjugated polymers by sol-gel-based in situ polymerization. Angewandte Chemie International Edition. 2001;**40**:3803-3806. DOI: 10.1002/1521-3757(20011015)113:20<3919:AID-ANGE3919>3.0.CO;2-5

[40] Lu Y, Yang Y, Sellinger A, Lu M, Huang J, Fan H, et al. Self-assembly of mesoscopically ordered chromatic polydiacetylene/silica nanocomposites. Nature. 2001;**410**:913-917. DOI: 10.1038/35073544

[41] Kimura M, Wada K, Ohta K, Hanabusa K, Shirai H, Kobayashi N. Organic-inorganic composites comprised of ordered stacks of amphiphilic molecular disks. Journal of the American Chemical Society. 2001;**123**:2438-2439. DOI: 10.1021/ja004034i

[42] Ikegame M, Tajima K, Aida T. Template synthesis of polypyrrole nanofibers insulated within one-dimensional silicate channels: Hexagonal versus lamellar for recombination of polarons into bipolarons. Angewandte Chemie International Edition. 2003;**42**:2154-2157. DOI: 10.1002/anie.200250800

[43] Li G, Bhosale S, Wang T, Zhang Y, Zhu H, Fuhrhop J-H. Gram-scale synthesis of submicrometer-long polythiophene wires in mesoporous silica matrices. Angewandte Chemie International Edition. 2003;**42**:3818-3821. DOI: 10.1002/anie.200351158

[44] Li G, Bhosale S, Bhosale S, Li F, Zhang Y, Guo R, et al. Template synthesis of functionalized polystyrene in ordered silicate channels. Chemical Communications. 2004:1760-1761. DOI: 10.1039/B402794A

[45] Clark AP-Z, Shen K-F, Rubin YF, Tolbert SH. An amphiphilic poly(phenylene ethynylene) as the

structure-directing agent for periodic nanoscale silica composite materials. Nano Letters. 2005;**5**:1647-1652. DOI: 10.1021/nl050472h

[46] Tu H-L, Tsai F-Y, Mou C-Y. One-step synthesis of ordered mesostructural organic/silica nanocomposites with tunable fluorescence surfactants. Journal of Materials Chemistry. 2008;**18**: 1771-1778. DOI: 10.1039/B800070K

[47] Okabe A, Fukushima T, Ariga K, Aida T. Color-tunable transparent mesoporous silica films: Immobilization of one-dimensional columnar charge-transfer assemblies in aligned silicate nanochannels. Angewandte Chemie International Edition. 2002;**41**: 3414-3417. DOI: 10.1002/1521-3773 (20020916)41:18<3414: AID-ANIE3414>3.0.CO;2-T

[48] Okabe A, Niki M, Fukushima T, Ariga K, Aida T. Ethanol vapor-mediated maturing for the enhancement of structural regularity of hexagonal mesoporous silica films. Chemical Communications. 2004:2572-2573. DOI: 10.1039/B410095F

[49] Wu J, Li J, Kolb U, Mullen K. A water-soluble hexa-*peri*-hexabenzocoronene: Synthesis, self-assembly and role as template for porous silica with aligned nanochannels. Chemical Communications. 2006: 48-50. DOI: 10.1039/B511868A

[50] Otani W, Kinbara K, Aida T. Template sol–gel synthesis of mesostructured silica composites using metal complexes bearing amphiphilic side chains: Immobilization of a polymeric Pt complex formed by a metallophilic interaction. Faraday Discussions. 2009;**143**:335-343. DOI: 10.1039/b904896k

[51] Lintang HO, Kinbara K, Tanaka K, Yamashita T, Aida T. Self-repair of a one-dimensional molecular assembly in mesoporous silica by a nanoscopic template effect. Angewandte Chemie International Edition. 2010;**49**:42411-44245. DOI: 10.1002/anie.200906578

[52] Lintang HO, Kinbara K, Yamashita T, Aida T. Heating effect of a one-dimensional molecular assembly on self-repairing capability in the nanoscopic channels of mesoporous silica. In: Proceedings of International Conferences on Enabling Science and Nanotechnology (ESciNano). Vol. 2010. Kuala Lumpur, Malaysia: IEEE; 2010. p. 5700970. DOI: 10.1109/ ESCINANO.2010.5700970

[53] Lintang HO, Kinbara K, Aida T. Thermally resistive phosphorescent molecular assembly in the channels of mesoporous silica nanocomposites. In: Proceedings of International Conferences on Enabling Science and Nanotechnology (ESciNano). Vol. 2012. Johor Bahru, Malaysia: IEEE; 2012. p. 6149684. DOI: 10.1109/ ESciNano.2012.6149684

[54] Lintang HO, Yuliati L, Endud S. Phosphorescent sensing and imaging of temperature using mesoporous silica/gold nanocomposites. Materials Research Innovations. 2014;**18**:S6 444-S6 448. DOI: 10.1179/1432891714Z.000000000995

[55] Lintang HO, Kinbara K, Yamashita T, Aida T. Metal-ion permeation in congested nanochannels: The exposure effect of Ag^+ ions on the phosphorescent properties of a gold(I)–pyrazolate complex that is confined in the nanoscopic channels of mesoporous silica. Chemistry--An Asian Journal. 2012;**7**:2068-2072. DOI: 10.1002/ asia.201200041

[56] Jalani MA, Yuliati L, Lintang HO. Thermal hydrogen reduction for synthesis of gold nanoparticles in the nanochannels of mesoporous silica composite. Jurnal Teknologi. 2014;**70**:131-136. DOI: 10.11113/ jt.v70.3420

[57] Jalani MA, Yuliati L, Endud S, Lintang HO. Synthesis of mesoporous silica nanocomposites for preparation of gold nanoparticles. Advanced Materials Research. 2014;**925**:233-237. DOI: 10.4028/www.scientific.net/AMR.925.233

Chapter 6

Mesoporous Materials for High-Performance Electrochemical Supercapacitors

Ranganatha Sudhakar

Abstract

Supercapacitors are very important kind of electrochemical energy storage devices. It differs from those of batteries and exhibit high power density. The energy storage in supercapacitors is influenced by many parameters like nature of electrode material, current collector, electrolyte, etc. among which most crucial is the morphology of the electrode. This makes most of the research works are targeting in designing suitable and high performing electrode materials by adopting new routes of synthesis, modifying the regular methods so on. Herein we discuss the fundamentals of supercapacitors, their design and approaches to obtain high-performance electrode materials.

Keywords: supercapacitors, porous, mesoporous, specific capacitance, carbon materials

1. Introduction

Materials with diverse structural and consequent physicochemical properties can be successfully evaluated for their electrochemical methods. Especially porous materials, which can be synthesized, modified or functionalized electrochemically or via other synthetic routes, exhibit good electrochemical properties. There is a direct link between porosity and electrochemistry and can be connected to following aspects like, composition and structural information by electroanalytical methods, electrosynthetic routes to design suitable porous materials, designing electrocatalysts for synthesis and sensing applications, magnetochemical, photochemical, electrochemical, photovoltaic etch [1–8] (**Figure 1**).

Generally there are many methods to analyze porous materials and electrochemical methods can be used to have analytical data on porous materials. Especially voltammetric techniques are extensively adopted as they provide information on reaction mechanisms. Also, AC impedance measurement aids to have information on corrosion and other metal surface related studies. This methodology, conceived as the recording of the voltammetric response of a solid material mechanically transferred to the surface of an inert electrode, provides information on the chemical composition, mineralogical composition, and specification of solids. Application of these porous materials especially intersecting with electrochemistry involves important concerns like transduction, sensing, gas production and its storage, energy production and storage and pollutant degradation. Application of porous materials is very significant in electrochemical energy generation and storage along with advanced dielectric materials for integrated circuits suitable for microelectronics industry.

Different electroanalytical techniques can be made use for getting information on the composition as well as structure of microporous materials. Generally, we can divide these techniques into two classes. Firstly, conventional electrochemical methods like cyclic voltammetry (CV), chronoamperometry, chronopotentiometry, and coulometry, and secondly, electrochemical impedance spectroscopy (EIS). We do have other additional methods namely, differential pulse- and square-wave voltammetries, electrochemical quartz crystal microbalance (EQCM), electrochemical atomic force microscopy which can efficiently characterize microporous solids. There is a possibility of having these techniques coupled with ultraviolet-visible spectrometry, Fourier-transform infrared spectroscopy, X-ray diffraction. Electrochemical phenomena involve electron transfer processes through a two-dimensional boundary (interface) separating thc mctal typc clcctrodc and thc ionically conducting electrolyte. Such kind of phenomena can be characterized by electrodics, focused on the heterogeneous electrode/electrolyte charge transfer process, and ionics, which throw light on ionically conducting liquid/solid phases. As far as porous materials are concerned, ionic conductivity being a general property, can be varied successfully by doping, defects and temperature. Astonishingly, some porous materials like hydrated aluminosilicates, behave as liquid electrolyte-like conductors, and as solid ionic conductors when dry. The classical model for describing the electrode-liquid electrolyte junction considers a formation of dense accumulation of oppositely charged species at the interface of electrode, double layer with dipole-oriented solvent molecules and a double layer of charge-separated ions, which results in a capacitive effect. Next to this, away from double layer, there is a less structured region, the diffuse layer, reducing randomly organized bulk-electrolyte solution. The earlier formulation, according to Helmholtz, distinguished between the inner (Helmholtz) layer, which composed all species that are specifically adsorbed on the electrode surface, and the outer (Helmholtz) layer, which comprises all ions closest to the electrode surface but are not specifically adsorbed. As area and geometry of the electrode surface affect

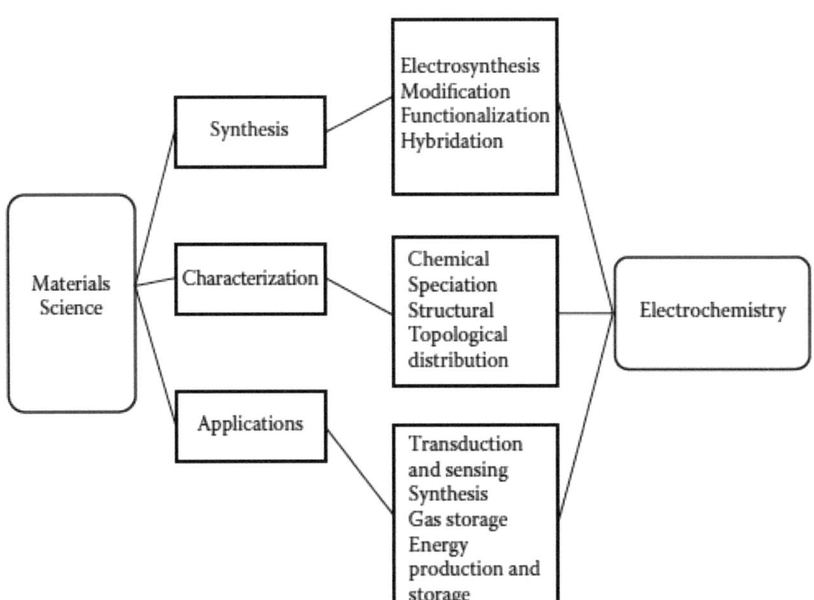

Figure 1.
Chart depicting link between diverse materials, their syntheses, characterizations and electrochemistry.

the double-layer capacitance, porous materials having large effective surface areas can lead to high capacitance effects, which will severely influence the electrochemical process. In the case when there is an establishment of some potential difference between electrode and electrolyte, some phenomena can be observed at the interface; charge transfer through electrode and electrolyte interface and concomitant charge transfer process in the electrolyte and electrode in particular involving ion restructuring in the double-layer zone. Resulting current flow can be described in terms of the summation of faradaic current which is associated with ion restructuring in the vicinity of the electrode surface. Let us consider an electroactive specie and its corresponding redox reaction or either a reduction or an oxidation reaction at an inert electrode surface. Faradaic current's magnitude is significantly influenced by mass transfer of these species from bulk of the solution to the surface of electrode. There are some particular ways how mass transport can occur. They are, diffusion where concentration difference makes transport happen, convection where momentum gradient drives and migration wherein electric field is the driving force. When solution is stirred or undergoes some vibrations, convection occurs. Ionic migration is diminished at highly concentrated supporting electrolyte. Under planar, semi-infinite diffusion conditions (vide infra), the faradaic current, i, for the reduction of a species whose concentration in the solution bulk is c, and its diffusion coefficient is D, at a plane electrode is then given by:

$$i = nFAD\left(\frac{\partial c}{\partial x}\right)_{x=0} \quad (1)$$

where A represents the electrode area, n is the number of transferred electrons per mole of electroactive species, and x is the distance from the electrode surface. This current is proportional to electroactive species concentration at the interface. Considering electron transfer phenomena across at the interface a heterogeneous one, heterogeneous electron transfer rate constant is applied to account the same.

The kinetics can be described via the Butler-Volmer equation:

$$i = -nFAk^0\left[C_{red}^0 \exp\left[\frac{nF(1-\alpha)\left(E - E^{0'}\right)}{RT}\right] - C_{ox}^0 \exp\left[-\frac{\alpha nF\left(E - E^{0'}\right)}{RT}\right]\right] \quad (2)$$

Here, concentrations of redox species are represented by C_{ox}^0 and $C_{red,}^0$ standard rate constant for the electron transfer (heterogeneous) corresponding to the reaction is k^0 and symmetry for the energy barrier is a. Also, E and $E^{0'}$ are applied and formal electrode potentials and $E - E^{0'}$ being overvoltage which represents the additional energy provided beyond equilibrium potential value. It is to be noted that, we obtain he Nernst equation from Butler-Volmer condition whenever system reaches equilibrium or current becomes zero and reaction will be very fast. Electrochemical reactions can be regarded as multistep as may be a reaction could be associated with reactions in solution phase or gas evolution or deposition. Electrochemical methods can provide mechanistic information of the processes by the aid of kinetics of interfacial reactions and coupled chemical reactions [9–15].

2. Mesoporous materials in energy storage and conversion

In current society, we are encountering energy crisis and consequently production, storage of electrical energy and its conversion to various other forms are most in must demand. Batteries are electrical power sources where chemical energy is converted into electrical energy via spontaneous electrochemical reactions. These power sources may be composed of one or several individual sets of electrodes making cells wherein reduction and oxidation reactions occur at specific electrode system which is separated by a diaphragm. These batteries are classified as primary and secondary which are characterized as chargeable and non-chargeable. There is a clear contrast between these batteries and capacitors which store electric energy with two conducting elements separated by a dielectric. With the application of potential difference between the two electrodes, capacitor acquires charge and stores. These capacitors in addition to be used in electronic circuitry, rectifiers, frequency filters, etc., can be also find application in energy storage especially for shorter periods. The charge/discharge processes in batteries goes with interconversions of electrode materials involving phase changes. Also, this process will be irreversible to some extent which restricts its cycle life which is a crucial parameter to decide whether battery is efficient or not. As achieving high stability and cyclability in energy storage devices along with the usage of environment friendly and inexpensive materials are the focus of considerable research efforts. In this context, the use of porous materials with high surface area plays an essential role.

Conventional capacitors are composed of nonpolarized dielectric materials such as ceramics or polymers, with picofarad to microfarad capacitance range. Polarized which are electrochemical capacitors involve generally anodic oxides of Al, Ta, or Nb, with the capacitance in microfarad to millifarad range. In recent past, new class of capacitors called supercapacitors are devised. These follow two probable mechanisms for charge storage: double layer and charge transfer reactions, the latter resulting in the so-called pseudocapacitances. Double-layer capacitance results from the separation of charge at the interface between a solid electrode and an electrolyte, whereas pseudocapacitance arises from fast, reversible faradaic reactions taking place at or near a solid electrode surface over an appropriate range of potential. Such redox reactions can go beyond the surface area and penetrate into the bulk of these materials. Metal oxides such as RuO_2 and IrO_2 offer the highest specific capacitance values (750 F/g), with the limitation that they are toxic and expensive. Hence, porous transition metal oxides such as CoO_x, NiO_x, and MnO_2 are currently under massive research as electrode materials for supercapacitors [16–20].

Electrochemical evaluation of their performance is usually conducted by preparing electrodes by pressing a mixture of the powdered material with a conducting material and binder i.e., acetylene black plus polytetrafluorene correspondingly, onto metal strip as current collectors (typically 1 × 1 cm^2). Cyclic voltammetric analysis of electrodes prepared from such materials allows researchers to differentiate between a purely capacitive behavior and a pseudocapacitance response. The purely capacitive response results in box-shaped voltammograms. The specific capacitance, Csp (F/g), can be evaluated from CVs as the quotient between the "box" current, I, and the product of the potential scan rate and the mass of the composite layer, m:

Analogously, the specific capacitance can be calculated from charge/discharge curves at a given current rate by using the relationship:

$$C_{xp} = \frac{I}{mv} \tag{3}$$

where I represents the discharge current density, m is the mass of the composite, Δt is the total time of discharge, and ΔV_d is the potential drop during discharge.

$$C_{xp} = \frac{I\,\Delta t}{m\,\Delta V_d} \tag{4}$$

The discovery of storing an electrical charge on surface arose from process associated with mechanical rubbing of amber in ancient times. Later in mid eighteen century the phenomenon was understood as "static electricity" and various "electrical machines" were being developed.

Scientists of General Electric Company were experimenting with devices using porous carbon electrode in 1957 and they found electric double layer capacitor effect. They observed carbon pores can store energy and it exhibited an exceptionally high capacitance. A research group in 1966 at Standard Oil of Ohio accidently rediscovered the same phenomenon while working on fuel cells. The cell was composed of two layers of activated charcoal sandwiched thin insulator, and the mechanical design remained the same for most electric double layer capacitors to date. In 1978 NEC introduce the term Supercapacitor and its application was used to supply backing up computer memory. Due to its application, many scientists delved into supercapacitor which led to trying of other composite as electrode material such as metal oxides and polymers especially conducting polymers, etc. Among the challenges faced in this century, energy storage is unquestionably significant. It is a challenge to search for environmentally friendly and low-cost energy storage material combination be found, in response to the needs of emerging ecological concerns and modern society. Supercapacitors are energy storage systems with very high capacity and a low internal resistance which are capable to store and deliver energy at relatively higher rates as compared to batteries due to the mechanism of energy storage which involves a simple charge separation at the interface between the electrode and the electrolyte. This is an alternate energy storage device with high power density and long cycle life. Supercapacitors are characterized with higher energy densities of about 5 Wh kg^{-1} in contrast to conventional capacitors. Although supercapacitors have the lower energy densities than secondary or chargeable batteries like Li ion batteries wherein their ultrafast charged and discharged capability lead to high power densities of about 10 kW kg^{-1}. The principle of energy storage in a supercapacitor is the ion adsorption on an electrode/electrolyte interface forming electric double layer leading to electrical double layer capacitors, EDLC or due to electron transfer between the electrolyte and electrode through fast Faradic redox reaction which is termed pseudocapacitors. These can find applications in electronics, transportation, communication and aviation. It can also be used in a wide range of energy capture and storage applications either by themselves as a primary power source or in combination with batteries and fuel cells. Supercapacitors can be used as back-up supplies used to protect against power disruption and load-levelers as back-up power for memories, microcomputers, clocks, system boards, etc. In the case of electric vehicles they are being used in combination with batteries and fuel cells. A supercapacitor consists of two electrodes, an electrolyte, and a separator which isolates the two electrodes electrically. Electrode material is the most important constituent of a supercapacitor. Key advantages of supercapacitors in comparison with other energy storage devices in terms of long life, high power, flexible packaging, wide thermal range of −40 to 70°C, low maintenance and low weight. Supercapacitors can best be made use in areas requiring applications with short load cycle as well as high reliability. Energy recapture sources such as forklifts, load cranes and electric vehicles, power quality improvement mostly utilize supercapacitors. Most promising application is in fuel cell vehicles and low emission hybrid vehicles. High power capability of supercapacitors can bridge the short time duration between a power failure and the startup of backup power generators. Though energy density of supercapacitor is greater than that of conventional capacitors which is considerably lower than batteries or fuel cells. Electrochemical performances of an electrode material strongly depend on parameters like surface area, electrical conductivity, wetting of electrode and permeability of electrolyte solutions which are very significant. The power needed by an application as well as the speed of storage process decides the kind of device to be applied. Indeed, when it comes to applications the ones that need faster discharge rate go for capacitor while the slower ones go for batteries. From **Figure 2**, it can be noted that batteries are capable of attaining up to 150 Wh/kg of energy density, around 10 times what an electrochemical supercapacitor is capable of. In terms of power density batteries do not possess the capability of having capacities of electrochemical capacitors. Batteries hardly reach 200 W/kg which is about 20 times less than that of expected electrochemical capacitor performance. Batteries experience shortcoming like rapid fall in performances due to fast charge discharge cycles or cold environmental temperature, they are expensive to maintain and have a limited cycle life [3–8].

2.1 Electrochemical supercapacitors

Based on the working principle, supercapacitors can be classified into three different groups. The first being electrical double layer capacitor EDLC where in capacitance is achieved due to pure electrostatic charge accumulates at the interface. This class of set up strongly depends on electrode surface area that is exposed to ions of electrolyte. Second is the pseudocapacitor, which considers a fast and reversible faradic process taking place due to electro active species. The third group is hybrid which combines the properties of both EDLC and pseudocapacitors. Selection of electrode material is of very high importance as it determines the resultant output. As charge storage in EDLC type is a surface phenomenon, surface property of electrode greatly affects capacitance of the capacitor. There are few challenges faced by supercapacitors include low energy density, production cost, low voltage per cell and high self-discharge. Most reliable way to overcome the shortcoming of low energy density is to design new electrode material. Popularly carbon materials, metal oxides and conducting polymers are used as electrode materials. Carbon materials have been used due to their high surface area. Metal oxides offer attractive choices as electrode material due to high specific capacitance with low resistance, making easier to construct high energy and power supercapacitors. Conducting polymers store and release charge based on reduction-oxidation processes. Supercapacitors cannot substitute batteries but could work as a supplement in terms of momentary and temporary power outage by providing instantaneous current required thereby reducing battery current. Electrochemical supercapacitors can be installed in parallel to compensate for momentary and

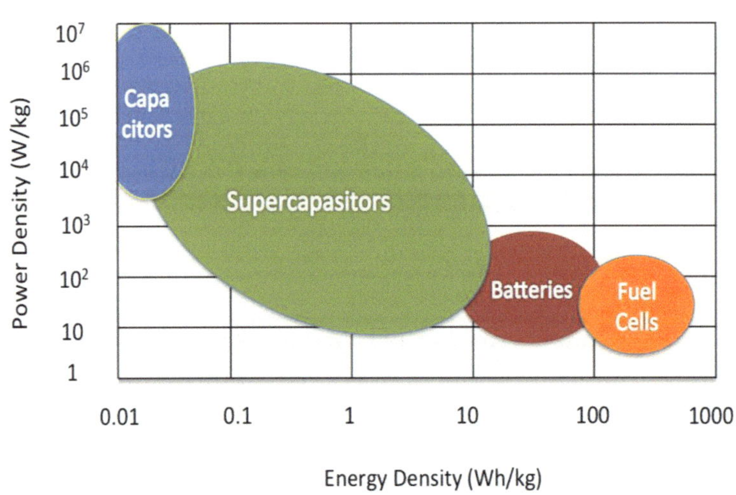

Figure 2.
Ragone plot showing energy density and power density for different kind of energy storage devices.

temporary interruptions in large scale battery units. Currently, all the researches focused on supercapacitors are on how to improve the energy density while maintaining high power density, fast charge/discharge and cycling stability.

Designing an EDLC involves two carbonaceous materials as electrodes, an electrolyte and a separator. The mechanism of charge storage in this case is through non-faradaic reactions that is via electrostatic one. There will be a creation of electrochemical double layer and voltage is applied there one can see accumulation of charges onto the porous surface of electrodes and ions in the electrolytic solution will diffuse towards opposite poles. The energy density at the electrode is proportional to specific surface area of the electrodes directly and their distance indirectly. Fast energy uptake, delivery and better power performance can be achieved by maintaining these parameters constructively. This principle of charge storage which is non-faradaic one is beneficial in contrast to that of batteries wherein electrode swelling is obvious because of charge and discharge processes. One can list out major differences between EDLC and battery. They are basically, wrt very long cycle life of EDLC compared to that of battery because of non-faradaic type of charge storage and non-involvement of solvent of electrolyte in charge storage (**Figure 3**).

Energy storage capacity and output of EDLCs strongly dependent on the type of electrolyte adopted. Reversible adsorption of ions from the electrolytes on the electrode surface drives the energy storage in double layer capacitors. Here there will be no electrochemical reactions and only physical absorption is considered and consequently no charge transfers hence no net ion exchanges across the interface of electrode and electrolyte. As strength of the electrolyte solution is constant and electrodes are intact throughout, this class of double layer capacitors are capable of sustaining millions of continuous charge/discharge cycles. The model of energy storage could be defined by the relation.

$$E = 0.5\,CV^2 \tag{5}$$

where E is the entire energy delivered, C is the capacitance and V is the working potential range [1–3].

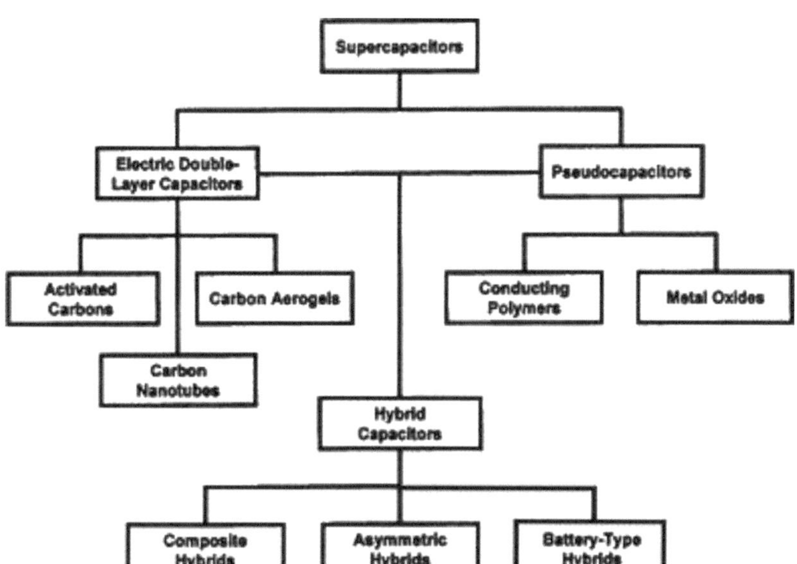

Figure 3.
Taxonomy of supercapacitors.

In contrast, charge storage in pseudocapacitors is dependent on faradaic process involving charge transfer at electrode electrolyte interface. Due to redox reactions between electrode and electrolyte when potential is applied makes to pass faradaic current through the supercapacitor cell. This makes them to exhibit higher specific capacitance compared to EDLCs. Frequently employed materials for this class faradaic materials are metal oxides, conducting polymers. As these involve in strong redox reactions as like battery materials, suffer lack of stability and cycle life. The pseudocapacitive class of capacitors, can be either surface phenomenon or bulk storage based on the location of the reactions. Pseudocapacitors involving multiple stable oxidation states, will be able to deliver higher energy density, however these suffer from relatively lower power density than EDLC types as Faradaic reactions are usually slower than those of non-faradaic processes. Also, it is to be noted that, because of redox reactions and corresponding physical changes in the electrodes, these show poor cycling stability compared to EDLCs.

The electrode materials mainly determine the performance of supercapacitors. Depending on the charge storage mechanism and active materials used, electrode materials for supercapacitors can be classified into three categories: (i) carbon materials, (ii) metal oxides/hydroxides and (iii) conducting polymers.

To compare and contrast EDLCs and pseudocapacitors, key parameters to be considered are cycle stability, energy and power densities. In the case of EDLCs, cycle stability and power performance will be very good, whereas, specific capacitance will be greater with pseudocapacitors. Hybrid systems combine benefits of both the mechanisms and offer extremely good power and energy characteristics, cycle life, cell voltage. Also, some advantages can be added, if one can play with different organic and inorganic electrolytes. Current research is focused on different configurations of these hybrid systems like composites, asymmetric and battery type (**Figures 4** and 5).

2.1.1 Mesoporous materials as electrode materials for supercapacitors

Research and innovations on complex and hierarchically organized porous materials has made tremendous progress and findings in the last decades and the area is still evolving. There has been many synthetic approaches developed towards designing porous materials that exhibit interconnected pore dimensions on several length scales, from molecular (0–2 nm) via nano- (2–100 nm) to macroscopic (4100 nm). Pores smaller than 2 nm are characterized micropores, pores with sizes between 2 and 50 nm mesopores and pores larger than 50 nm are macropores. Such multilevel porous architectures confer unique properties and features to materials depending on the combination of pore sizes, e.g., micro- and mesopores impart high surface areas and pore volumes providing size and shape selectivity and large interfacial areas, while larger pores of about 450 nm reduce transport limitations in the material and facilitate mass transport to the active sites. Diverse techniques and routes have already been reported for the preparation of micro-macroporous, micro-mesoporous, meso-macroporous or micro-meso-macroporous materials with great potential for applications in the fields of catalysis, sorption, separation, energy storage and conversion, sensing and biomedicine, i.e., medical diagnostics or therapies. Structural characteristics such as the total amount of pores, the accessibility of the pores (ratio of closed to open pores), tortuosity and inter connectivity, gradients, etc., and very importantly, the chemical composition as well as the processability in terms of shaping (films, fibers, monoliths, etc.) make materials suitable for specific applications. As an example, for the importance of structure and morphology of the materials, highly porous, hierarchically organized macroscopic monolithic material, can give lower backpressures, a higher permeability and better performance in flow-through catalytic or separation systems [13–18].

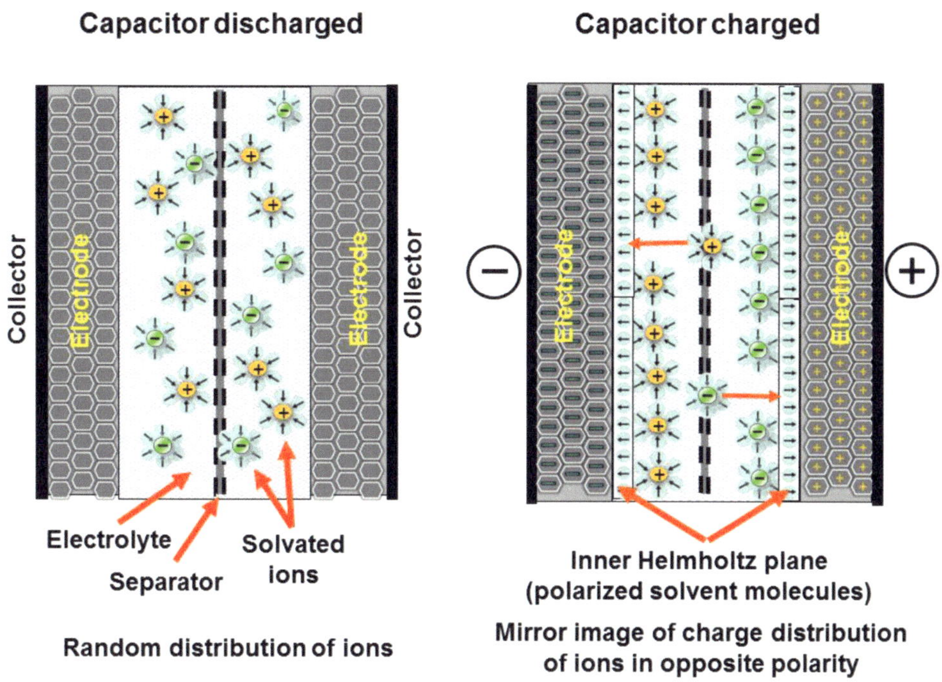

Figure 4.
Mechanism of charge storage in electrochemical double layer capacitor (EDLC).

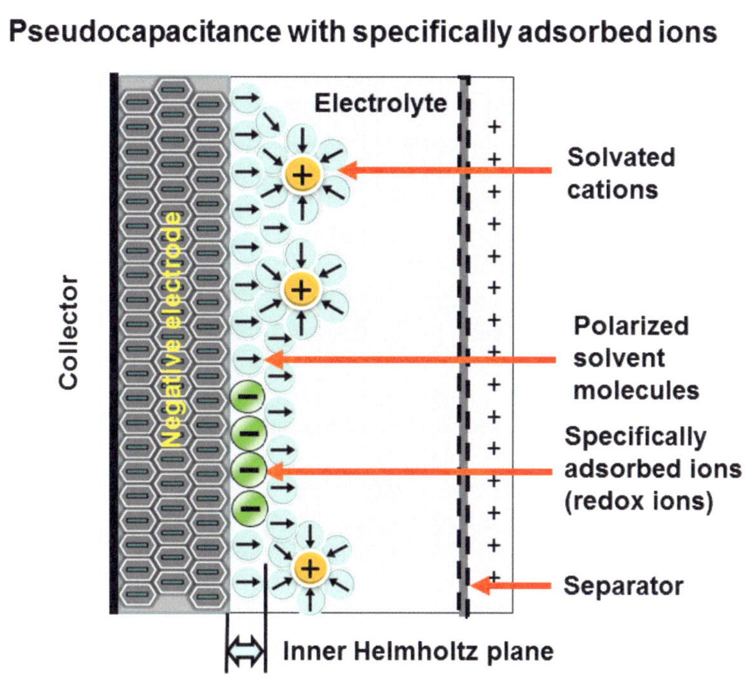

Figure 5.
Mechanism of charge storage in pseudocapacitors.

Materials with high surface areas are considered to enhance supercapacitor performance. In addition, pores in the electrode material not only increases the surface area of the material but also facilitates charge to be stored in these micro/meso pores. Electrodes in EDLCs have been designed using many types of carbonaceous materials including activated carbons, carbon nanotubes (CNTs), carbon aerogel (CAGs), and carbon nanofibers (CNFs) and graphene. Activated carbon possesses very high surface area, relatively low cost and largely available and therefore a popular electrode material. It exhibits differently porous structures and classified based on their pore sizes like micro (<2 nm), meso (2–50 nm), macro (>50 nm) associated with high surface area. Ions get adsorbed and accumulated in the micropores to form double layer and ion diffusion to these micropores will be via mesopores which offers low resistive paths for electrolytic ions. These pores can be controlled to have uniform pore sizes with highly ordered nanochannels and large surface area in template carbons. Templated carbonization is superior than activated carbon in terms of degree of control over pore size distribution and interconnection of pores. Charge storage and rate capability can be seen enhancing in the case of interconnected pore structures. Alternatively, CNTs are also a common class of electrode materials as it is mechanically strong and highly conducting. These show higher accessibility to electrolytic ions due to mesopores of CNTs which are larger than micropores of activated carbons. One more class of carbonaceous materials, CAGs also being employed as electrode materials as they possess outstandingly high specific surface area and significant electrical conductivity. Mostly sol–gel method will be employed to design these CAGs, using resorcinol and formaldehyde aerogels. In recent past, graphene has become most popular materials for supercapacitor due its chemical stability, large surface area and high conductivity as a carbon based material, with restacking of sheets being a problem while fabricating the electrodes. CNFs which are derived from carbon based polymers are known for greater mechanical strength, surface area, flexibility and conductivity. The flexibility and freestanding capacity of CNFs dismisses the need for using binders during fabricating electrodes in turn enhancing the mass fraction of the active material in the device. Electrospinning is a simple and widely employed technique to synthesize nanofibers from a wide group of polymers. In general, the polymer blend technique is a promising route to design carbon materials with fine and controlled pores. This can be achieved by blending of two immiscible polymers allowing control over the blend texture to obtain desired pore sizes.

Supercapacitors have attracted a lot of attention recently as energy storage devices, due to their high power density, long cyclability, wide working potential window and broad temperature range. Though supercapacitors offer a higher energy density, they possess lower energy density compared to those of batteries. Therefore, there has been a lot of scope improving energy density of supercapacitors high power density. As mechanism of charge storage is different in different types of supercapacitors, routes to improvise energy density also varies. As EDLCs store the charges electrostatically, high surface area and porosity of the electrode material enhance the energy density. Additionally, type of the material, like CDC, graphene, CNF, etc. further influence the extent of charge storage in turn the energy. Among these, carbon nanofibers are one of the most promising candidates as they pose relatively good mechanical strength, high surface area, relatively higher conductivity, freestanding nature and flexibility.

One of the most popularly employed electrode material is activated carbon (AC) which may be attributed to large surface area, good electrical properties and moderately economical. AC can be prepared by either physical or chemical activation from various types of carbonaceous materials like wood, coal nutshell, etc. These can be physically activated at temperatures 700–1200°C in the presence of oxidizing gases like CO_2 and air. Chemical activation is carried out at a lower temperature 400–700°C employing sodium hydroxide, potassium hydroxide, zinc chloride and phosphoric acid. Activation process imparts AC to possess numerous physiochemical properties with well-developed surface areas of up to 3000 m^2/g. Porous structure of AC obtained after activation processes will have a broad pore size distribution consisting of micropores (50 nm). Research efforts have been made to see the effect of different electrolytes on the capacitance performance of AC. It was observed that the capacitance of AC is higher in aqueous electrolytes (ranging from 100 to 300 F/g) as compared to organic electrolytes [10–14].

With the discovery of carbon nanotube (CNT) there has been a significant advancement in the science and engineering of carbon materials. The factor that determines the power density in a supercapacitor is overall resistance of the components. Seriously attention is been paid to CNT as supercapacitor material because of its unique pore structure, good mechanical and thermal stability with superior electrical properties. Catalytic decomposition is the most frequently adopted method to synthesize carbon nanotubes and further careful manipulation of different parameters, it becomes possible to obtain nano structures in various conformations and also control their crystalline structure. Carbon nanotube have interconnected mesopores which allow charge distribution that utilizes almost all accessible surface area. These have a lower ESR than those of activated carbon as the electrolyte ions can diffuse into the mesoporous network. CNT can be typically of two types; single-walled carbon nanotubes (SWCNTs) or multi-walled carbon nanotubes (MWCNTs), both of which can be employed as supercapacitor electrode materials. While targeting high power electrode materials, CNT are considered due to their remarkable electrical conductivity and readily accessible surface area. In addition, they offer support for active materials due to their high mechanical strength and open tubular network structure [12–16].

Graphene has enjoyed significant attention from scientific community. Graphene a one atom thick layer 2D structure has evolved as a unique and special carbon material that has capacity to store energy because of its superior characteristics of high large surface area, electrical conductivity and chemical stability. Graphene has been recognized as superior supercapacitor electrode material as the charge storage is not dependent on pore distribution unlike in the case of other carbonaceous materials. Specific surface area of about 2500 m^2/g could be obtained and if it is utilized completely, capacitance of 500–600 F/g could be achieved. Following methods can be employed to produce graphenes, like micromechanical exfoliation, chemical vapor deposition, unzipping of CNTs, epitaxial growth, arch discharge method, electrochemical and chemical methods and intercalation methods in graphite. Many strategies are being applied so as to refrain restacking of graphene sheets so that high specific surface area and intrinsic surface capacitance could be achieved. To list out major reports from literature, thermally exfoliated graphitic oxide offers specific capacitance of 117 F/g and energy density of 31.9 Wh/kg. Graphene from modified hummer's method and tip sonication, energy and power density obtained were 58.25 Wh/kg and 13.12 kW/kg at 7.5 A/g, which makes suitable for electric vehicle. Thermal reduction of graphite oxide at elevated temperature results in highly corrugated graphene exhibiting specific capacitance of 349 F/g [17–20].

3. Conclusion

Thus porosity of the material is very crucial and significant property in deciding performance in electrochemical energy storage. Different strategies are being practiced to obtain a porosity in turn a high surface area. The improved porosity and high surface area provides larger interfacial area between electrode and electrolyte which provides excess of interaction and charge storage.

Author details

Ranganatha Sudhakar
Department of Chemistry, School of Engineering, Presidency University, Bengaluru, India

*Address all correspondence to: kamath.ranganath@gmail.com

References

[1] Conway BE. Electrochemical Supercapacitors: Scientific Fundamentals and Technological Applications. New York: Kluwer Academic/Plenum Publishers; 1999

[2] Arico AS, Bruce PG, Scrosati B, Tarascon JM, Schalkwijk WV. Nanostructured materials for advanced energy conversion and storage devices. Nature Materials. 2005;**4**:366-377

[3] Simon P, Gogotsi Y. Materials for electrochemical capacitors. Nature Materials. 2008;**7**:845-854

[4] Hall PJ, Mirzaeian M, Fletcher SI, Sillars FB, Rennie AJR, Shitta GOB, et al. Energy storage in electrochemical capacitors: Designing functional materials to improve performance. Energy & Environmental Science. 2010;**3**:1238-1251

[5] Wang G, Zhang L, Zhang J. A review of electrode materials for electrochemical supercapacitors. Chemical Society Reviews. 2012;**41**:797-827

[6] Hou J, Shao Y, Ellis WM, Obert B, Moore BY. Graphene based electrochemical energy conversion and storage: Fuel cells, supercapacitors and lithium ion batteries. Physical Chemistry Chemical Physics;**2012**(13):15384-15402

[7] Patil S, Raut S, Gore R, Sankapal B. One dimensional cadmium hydroxide nanowires towards electrochemical supercapacitors. New Journal of Chemistry. 2015;**39**:9124-9131

[8] Ma X, Zhang W, Kong L, Luo Y, Kang L. $NiMoO_4$-modified MnO_2 hybrid nanostructures on nickel foam: Electrochemical performance and supercapacitor applications. New Journal of Chemistry. 2015;**39**:6207-6215

[9] Hu CC, Chang KH, Lin MC, Wu YT. Designing and tailoring of the nanotubular arrayed architecture of hydrous RuO_2 for next generation supercapacitor. Nano Letters. 2006;**6**:2690-2695

[10] Brousse T, Belanger D, Long JW. To be or not to be pseudocapacitive? Journal of the Electrochemical Society. 2015;**162**(5):A5185-A5189

[11] Zhao C, Zheng W. A review for aqueous electrochemical supercapacitors. Frontiers in Energy Research. 2015;**3**:23

[12] Dubal DP, Ayyad O, Ruiz V, Gomez-Romero P. Hybrid energy storage: The merging of battery and supercapacitor chemistries. Chemical Society Reviews. 2015;**44**:1777-1790

[13] Chen K, Xue D. Colloidal supercapacitor electrode materials. Materials Research Bulletin. 2016;**83**:201-206

[14] Mohmood N, Zhang C, Yin H, Hou Y. Graphene based nanocomposites for energy storage and conversion in lithium batteries, supercapacitors and fuel cells. Journal of Materials Chemistry A. 2014;**2**:15-32

[15] Boenstein A, Hanna O, Attias R, Luski S, Brousse T, Arbach D. Carbon based composite materials for supercapacitor electrodes: A review. Journal of Materials Chemistry A. 2017;**5**:12653-12672

[16] Pumera M, Sofer Z, Ambrosi A. Layered transition metal dichalcogenides for electrochemical energy generation and storage. Journal of Materials Chemistry A. 2014;**2**:8981-8987

[17] Li B, Zheng M, Xue H, Pang H. High performance electrochemical

capacitor materials focusing on nickel based materials. Inorganic Chemistry Frontiers. 2016;**3**:175-202

[18] Wang Y, Song Y, Xia Y. Electrochemical capacitors: Mechanism, materials, systems, characterization and applications. Chemical Society Reviews. 2016;**45**:5925-5950

[19] Feinle A, Elsaesser MS, Husing N. Sol-gel synthesis of monolithic materials with hierarchical porosity. Chemical Society Reviews. 2016;**45**:3377-3399

[20] Yang Q, Lu Z, Liu J, Lei X, Chang Z, Luo L, et al. Metal oxide and hydroxide nanoarrays: Hydrothermal synthesis and applications as supercapacitors and nanocatalysts. Progress in Natural Science: Materials International. 2013;**23**(4):351-366